Experiments in General, Organic, and Biological Chemistry

Experiments in General, Organic, and Biological
CHEMISTRY

Fourth Edition

Robert J. Ouellette
The Ohio State University

Jason H. Manchester
The Ohio State University

Prentice Hall

PRENTICE HALL Upper Saddle River, NJ 07458

Acquisitions Editor: **Ben Roberts**
Production Editor: **Kim Dellas**
Special Projects Manager: **Barbara A. Murray**
Supplement Editor: **Ashley Scattergood**
Production Coordinator: **Ben Smith**

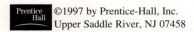 ©1997 by Prentice-Hall, Inc.
Upper Saddle River, NJ 07458

Printed in the United States of America

15   14   13

ISBN 0-13-286758-3

Prentice-Hall International (UK) Limited, *London*
Prentice-Hall of Australia Pty. Limited, *Sydney*
Prentice-Hall Canada Inc., *Toronto*
Prentice-Hall Hispanoamericana, S.A., *Mexico*
Prentice-Hall of India Private Limited, *New Delhi*
Prentice-Hall of Japan, Inc., *Tokyo*
Prentice-Hall Asia Pte. Ltd., *Singapore*
Editora Prentice-Hall do Brasil, Ltda., *Rio de Janeiro*

Preface

This laboratory manual has been designed for use in the course taken by students of nursing, home economics, agriculture, and the health sciences. It is intended for a course that includes the fundamentals of inorganic, organic, and biological chemistry.

The manual is written with the assumption that the students have had no previous chemistry laboratory experience. The objectives are

1. To introduce students to simple laboratory operations.
2. To increase their familiarity with chemical reactions discussed in the textbook.
3. To develop a sense of the limitations of measurement.
4. To develop methods of displaying numerical data in tabular form as well as in graphical form.
5. To teach how to draw conclusions from data.

Laboratory techniques are discussed in a separate section. This placement allows experiments to be done in any order. Instructions in each experiment refer to the appropriate techniques section.

Safety precautions are highlighted in each experiment. Only by observing these precautions can you and your fellow students be safe in the laboratory.

All reagents and necessary equipment are listed at the beginning of each experiment. An introduction to the experiment is given in sufficient detail to allow students to work with a minimum of supervision.

Most of the experiments in this manual have been worked, in many revised forms, over a period of 30 years by many students at The Ohio State University. Each revision has been evaluated by feedback from both students and teachers. Seven new experiments have been added to this edition. The authors would appreciate learning of any errors in the production of this edition of the manual.

R.J.O.
J.H.M.

Contents

Safety in the Laboratory

The chemistry laboratory and the experiments to be done have been designed to make you and your classmates as safe as possible. There is no reason to have accidents in the laboratory. Whether accidents occur or not is largely your responsibility. Only by reading all instructions and then following them exactly can you prevent accidents. It is mandatory that you conduct your experiments safely, for others could be injured as a result of your carelessness.

Accidents in the laboratory can result from

1. Failing to read all instructions carefully.
2. Not listening to all precautions of the instructor.
3. Doing unauthorized experiments.
4. Using incorrect experimental techniques.

The following safety rules and procedures must be obeyed. If you are not willing to contribute to a safe environment in the laboratory, you will not be allowed to continue in the course.

1. The most important safety precaution in the laboratory is the wearing of approved safety goggles. Ordinary prescription glasses, contact lenses, or safety glasses are not adequate. Safety goggles provide protection from both the impact of projectiles and the splashing of chemicals. Most states have a law mandating that such eye protection be worn.

    Put on your safety goggles immediately when you open your laboratory locker.

    As you prepare to leave at the end of the laboratory period, take off your safety goggles only after everything has been cleaned and put away. Exit the laboratory promptly.

2. It is wise not to wear your best clothes to the laboratory. Your clothing should be as protective and safe as possible. The clothing layer can protect your skin from contact with chemicals. For this reason shorts are not advisable. Shoes should be worn to protect your feet. Loose or large sleeves should be avoided, since they may accidentally drop into chemicals. Hair should be tied back.

3. Do not start the experiment until the instructor arrives. **Never work in the laboratory alone.**

4. Know the locations of the eyewash fountains, fire blanket, and safety shower.

5. Know the locations of fire extinguishers, fire alarms, and emergency phones.

6. Never taste any chemicals in the laboratory. All substances should be treated as toxic.

7. Smelling a solid, liquid, or gas should be done *only* if required in the experiment. Waft the vapors toward your nose with your hand. Do not breathe a chemical directly.

8. Do not eat or drink in the laboratory. You can accidentally ingest chemicals with the food. Wash your hands after you leave the laboratory.

9. Do not smoke in the laboratory. Someone else may be working with flammable material. A fire or explosion can result even if the flammable material is at some distance.

10. Do not use a match or laboratory burner in an experiment with flammable liquid in the vicinity. Do not leave a burner unattended.

11. When heating liquid in a test tube, do not point the open end toward yourself or anyone else.

12. Hot glass looks the same as cold glass. Put hot glass out of the way and remember not to touch it until it has cooled.

13. Read all labels carefully so that you use the proper chemical for an experiment. Never return excess unused chemical to a reagent bottle.

14. Never place the stopper from a reagent bottle on the laboratory bench. The chemical may become contaminated when the stopper is replaced. The bench will contain residual chemicals that someone else may touch.

15. Dispose of liquid and solid wastes exactly as directed. Do not flush chemicals down the drain. Certain flammable solvents and other chemicals are to be placed in specially marked containers.

16. Dilute an acid only by slowly pouring it into water with stirring. Never pour water into acid.

17. Use the fume hood when working with toxic or irritating chemicals.

18. Keep a neat, safe laboratory by cleaning up spills and broken glass immediately.

19. Know the emergency procedures of your laboratory so that you can act promptly if an accident does occur to you or a classmate.

20. Report all accidents, no matter how small, to your instructor immediately.

Laboratory Techniques

Chemistry is an experimental science. The chemical knowledge obtained by laboratory experiments in the past two centuries is the basis of today's chemistry courses. Some of this knowledge was obtained by experiments carefully designed to observe the behavior of matter; some derived from observations that were entirely unanticipated. In either case, through the careful recording of qualitative and quantitative observations the laws of chemistry have been developed.

In order to understand and appreciate the science of chemistry, one must do experiments. Although the scientific method is described in most textbooks, the reasoning process is better appreciated by doing laboratory work. Experimental difficulties often are encountered in the laboratory; care and patience are required to overcome them. Obtaining accurate and reproducible data contributes more than any textbook to understanding the relationship between experimental observations and the development of laws and theories.

Ideally, the experiments in the laboratory should correspond to the material being discussed in the lecture. However, such timing is not always possible. This problem should encourage you to do what every scientist strives to do. You should plan your experiment before going to the laboratory. **Read the experiment first.** Think about what you are to do. Refer to your textbook so that you at least roughly understand the necessary background material. Your instructor can help you with your difficulties in an experiment if you have prepared adequately in advance. The significance of the experiment and the results will be most meaningful to the prepared student.

Chemists have developed many techniques for conducting safe and successful experiments. Your assigned work depends on your understanding these techniques. Study and review the appropriate techniques before starting each new laboratory project. The equipment to be used in this course is shown on the next three pages.

The techniques discussed in this section include

A.	Handling Liquid Reagents	I.	Reading a Meniscus
B.	Handling Solid Reagents	J.	Volumetric Analysis
C.	Testing for Odors (Wafting)	K.	Melting Point
D.	Filtration	L.	Spectrophotometry
E.	The Gas Burner	M.	Use of the Laboratory Barometer
F.	Heating Liquids	N.	Use of the Balances
G.	Use of the Crucible	O.	Use of the Digital Thermometer
H.	Glassworking	P.	Graphing

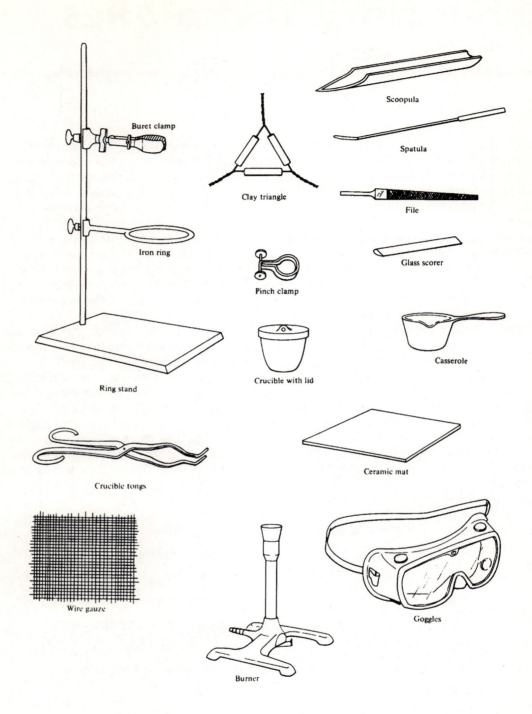

Buret clamp

Iron ring

Ring stand

Clay triangle

Pinch clamp

Crucible with lid

Scoopula

Spatula

File

Glass scorer

Casserole

Crucible tongs

Ceramic mat

Wire gauze

Burner

Goggles

COMMON LABORATORY EQUIPMENT (continued)

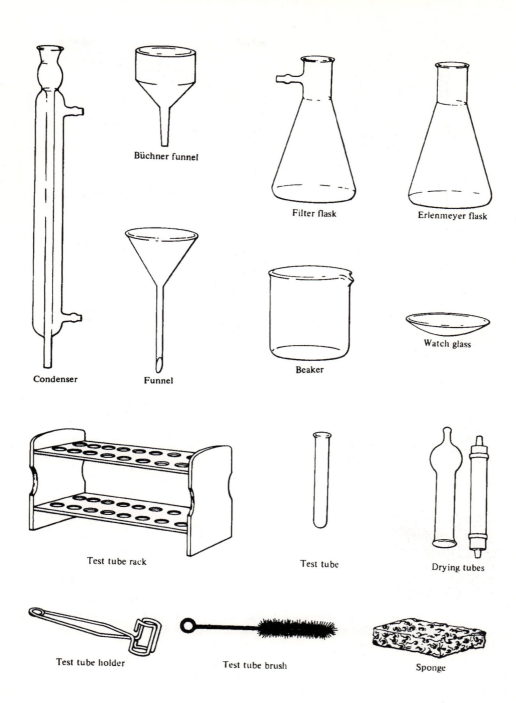

Büchner funnel

Filter flask

Erlenmeyer flask

Condenser

Funnel

Beaker

Watch glass

Test tube rack

Test tube

Drying tubes

Test tube holder

Test tube brush

Sponge

COMMON LABORATORY EQUIPMENT (continued)

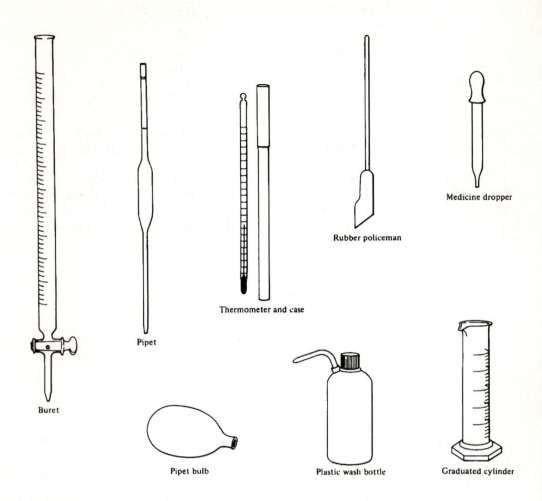

Buret

Pipet

Thermometer and case

Rubber policeman

Medicine dropper

Pipet bulb

Plastic wash bottle

Graduated cylinder

A. HANDLING LIQUID REAGENTS

First check the label to be certain that you have the correct reagent. Tilt the bottle until the liquid wets the stopper. Now hold the bottle at a slight angle and twist the stopper to loosen it (Figure 1). Place the stopper firmly between the fingers of the hand used to hold the bottle as shown in Figure 2. **Do not lay the stopper on the bench.** Impurities may be picked up and transferred to the reagent when the stopper is returned to the bottle.

By holding the stopper and bottle in the same hand, your other hand is free to hold another vessel. Now pour the liquid into the receiving vessel. A stirring rod can be used to avoid splashing (Figure 2). Try not to take more reagent than needed. **Never pour unused reagents back into the stock bottle.** This avoids unintentional contamination. Excess reagent should be given to another student or discarded properly. Do not leave a reagent bottle unstoppered.

| Figure 1. Removing a glass stopper from a reagent bottle. | Figure 2. Pouring a liquid from a reagent bottle. |

For a rough determination of volume, a liquid reagent is poured into a graduated cylinder. A measured volume is then poured from the graduated cylinder into another vessel. Note that the graduated cylinder may have a plastic collar (Figure 3). This collar should be positioned somewhere above the highest graduation. If the graduated cylinder is tipped over accidentally, the collar will prevent breakage.

Figure 3. Graduated cylinder with a plastic safety collar.

Disposal of Liquid Reagents

Compounds are divided into two classes: organic and inorganic. Many inorganic reagents are used as aqueous solutions. The proper disposal of these reagents varies considerably. Your instructor will inform you of the procedure required by State and Federal regulations.

Organic liquids pose two problems for disposal. First, many are flammable; second, most do not mix with water. For these reasons, do not pour organic liquids down the drain. Pour waste organic liquids into the special flameproof can provided in the laboratory room. This can is bright red and is labeled organic waste. Only organic waste is to be put in this can. Consult your instructor if you have any questions about the proper disposal of any liquid.

B. HANDLING SOLID REAGENTS

Read the label on the bottle to be sure that you are using the correct reagent. The bottle may have a glass stopper, a rubber stopper, or a screw cap. Hold the bottle firmly on the bench and remove the stopper or screw cap. Place the screw cap or stopper on the bench inner side up. Hold the bottle with the label against your hand and then tilt and rock the bottle back and forth as shown in

Figure 4. The solid should be transferred into a receiving vessel. Avoid spilling any solid reagent on the bench. Ask your instructor how to clean up solid reagent if some is spilled. **Recap the bottle.** Try not to take more reagent than you need. **Do not return excess reagent to the stock bottle.** Give the excess reagent to another student or dispose of it properly.

Label against the hand

Roll and pour

Figure 4. Dispensing a solid reagent.

Disposal of Excess Solid Reagents

Many inorganic solids are soluble in water. Some may be flushed down the drain with large amounts of water. Your instructor will inform you of any special procedures prescribed by State and Federal standards. Insoluble solids should be placed in the solid waste can. Consult your instructor for other special procedures for solid waste disposal.

C. TESTING FOR ODORS (WAFTING)

You will test for chemical odors in several experiments. Some of these odors are strong and/or hazardous. Others are weak and/or harmless. Always assume that a chemical odor is strong and hazardous. **Do not place your nose directly over the vessel. Do not inhale deeply.**

The correct way to smell chemicals is by wafting (Figure 5). Bring the top of the vessel near your nose and use your free hand to gently fan (waft) the vapors toward your nose. Inhale slowly and cautiously. If no odor is detected, then somewhat more vigorous wafting may be attempted.

Figure 5. Testing for odors. Fan (waft) the vapors gently toward the nose.

D. FILTRATION

This technique uses filter paper to separate a solid from a liquid. Filter paper is porous and allows liquids but not solids to pass through it. You will use two filtration techniques: gravity and suction.

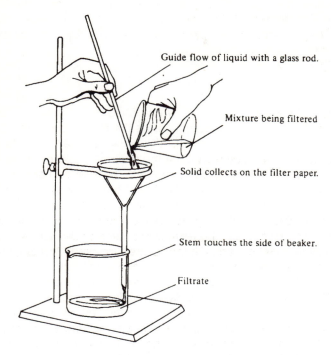

Guide flow of liquid with a glass rod.

Mixture being filtered

Solid collects on the filter paper.

Stem touches the side of beaker.

Filtrate

Figure 6. Apparatus for gravity filtration.

Gravity Filtration

An ordinary funnel is used for gravity filtration (Figure 6).   A circular piece of filter paper is folded to fit snugly in the funnel.   The folding procedure is shown in Figure 7.

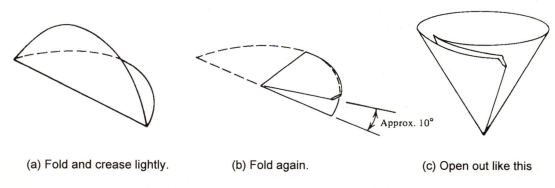

Approx. 10°

(a) Fold and crease lightly.   (b) Fold again.   (c) Open out like this

Figure 7. Sequence of folding filter paper for gravity filtration.

D. Filtration

Note that a small part of the corner of the folded paper is torn off to help seal the wet paper against the edge of the funnel.

Support the funnel with a buret clamp or funnel support. Place the folded filter paper in the funnel (Figure 6). Moisten the paper with the solvent to be filtered. Seal the filter paper against the funnel by pressing the top of the moist paper against the funnel.

Filter the solid as shown in Figure 6. The tip of the funnel should touch the wall of the beaker to avoid loss due to splashing. Do not fill the funnel more than two-thirds full. Try to keep the stem filled with the filtrate (the liquid passing through the filter paper) because its weight causes a slight suction that accelerates filtration.

Suction Filtration

A typical setup for suction filtration is shown in Figure 8. The Büchner funnel used in suction filtration has a flat bottom with small holes. This flat bottom supports the filter paper when suction is applied. Suction is provided by a water aspirator (Figure 9). The aspirator is attached to the cold water faucet at the sink. When water flows rapidly through the aspirator, suction is created in the side arm. Connection to the filter trap is made by a rubber hose to the side arm of the aspirator. The trap is provided to catch any water that might back up from the aspirator. Without the trap, any back-up water would go into the filter flask and contaminate the filtrate.

To use the apparatus, place a piece of filter paper in the Büchner funnel. Moisten the paper with a small amount of the solvent to be filtered. Seal the paper by turning on the aspirator to create a small vacuum. Carefully add the material to be filtered to the funnel and increase the suction from the aspirator. Do not fill the Büchner funnel more than two-thirds full.

Wash the remaining solid into the Büchner funnel with pure solvent. Use several portions of 2-5 mL each. After all of the liquid has drained into the filter flask, break the vacuum. The vacuum is broken by slowly pulling the hose off the aspirator while maintaining the flow of water through the aspirator. If the flow of water is stopped prior to disconnecting the hose, tap water may back up into the filtration apparatus.

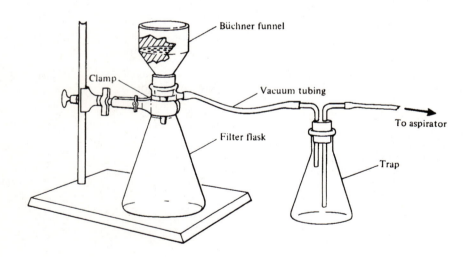

Figure 8. Apparatus for suction filtration.

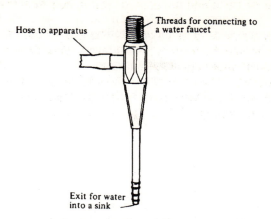

Figure 9. Water aspirator for suction filtration.

Remove the Büchner funnel and invert it on a piece of clean white paper. Scrape the solid from the filter paper using a spatula. Fold the paper and use it to slide the solid into a small beaker, a vial, or a plastic bag. The filtrate may be poured directly from the filter flask into any other vessel.

E. THE GAS BURNER

The gas burner is a device widely used in chemical laboratories. Since the first laboratory burner was perfected by Robert Bunsen, many people give his name to any simple laboratory burner. Laboratory burners are made in many shapes and sizes, but all of them accomplish the same purpose. The burner forms a gas-air mixture that will burn to give a hot, efficient flame. A Meker burner is shown in Figure 10.

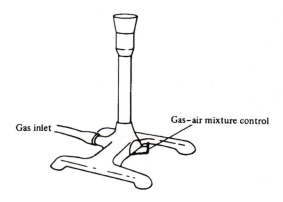

Figure 10. Meker burner.

The natural gas available in most laboratories consists mainly of methane. When mixed with sufficient oxygen of the air, methane will burn to give carbon dioxide and water. The heat given off by this exothermic chemical reaction is used to heat liquids and solids in laboratory apparatus.

To use the burner, connect one end of a rubber hose to the gas inlet on the base of the burner. Attach the other end to the gas jet on the bench. Slowly open the gas valve and bring a lighted match to the top of the burner. Adjust the flame by moving the gas-air mixture control on the bottom of the burner. To obtain the hottest flame, open the gas valve all the way, and then open the mixture control until the burner starts to produce a hissing sound.

For a Meker burner, the flame should be about 2 cm high. The hottest position in the flame is about 1 cm above the top of the burner. The hottest part of the flame is just above the tip of the blue part of the flame (Figure 11).

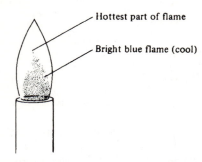

Hottest part of flame

Bright blue flame (cool)

Figure 11. Flame of a properly adjusted burner.

F. HEATING LIQUIDS

Small amounts of liquids may be heated for a short time in a test tube. Larger amounts of liquids are heated in a flask or beaker. Heating a large amount of a liquid for a prolonged period is done under reflux.

Heating in a Test Tube

Fill the test tube not more than one-third full and position it in the flame as shown in Figure 12. The flame should heat at the top of the liquid, not at the bottom.

Position flame near leading edge of liquid to avoid "bumping."

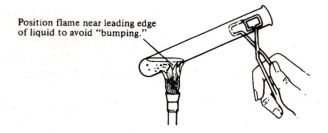

Figure 12. Heating a test tube with a burner.

Improper heating may cause a violent ejection of the liquid from the test tube. For this reason, aim the mouth of the test tube in a safe direction. **Never point the mouth of the test tube toward your neighbor or yourself.**

Heating in a Flask or Beaker

Attach an iron ring to a ring stand. Place a wire gauze between the beaker or flask and the iron ring as shown in Figure 13. Turn on the gas, light the burner, and place it under the iron ring.

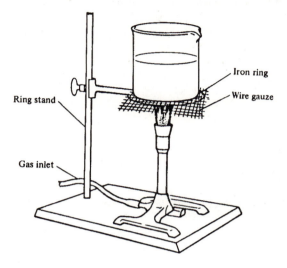

Figure 13. Heating a liquid in a beaker.

Refluxing a Liquid

A liquid may need to be heated for an extended time. However, evaporation of the liquid would then occur. To avoid evaporation, a condenser is attached to the flask that contains the liquid (Figure 14). Cold water is run through the outside of the condenser. The hot vapors from the liquid are cooled inside the condenser and form the liquid again. The condensed liquid then flows back into the flask and is reheated. The process of constantly boiling and condensing a liquid is called refluxing.

Often boiling occurs very roughly and unevenly. Bumping is caused by the sudden formation of superheated vapor near the flame. Bumping can be avoided by the addition of boiling chips (boiling stones). A boiling chip is a small white porous stone that facilitates smooth boiling on its outer surface.

To reflux a liquid, assemble the apparatus as shown in Figure 14. Place the liquid to be refluxed in the Erlenmeyer flask and add about five boiling chips. Connect the lower water hose of the condenser to a cold water faucet. Place the hose from the top of the condenser in a drain. **Be sure all hose connections are secure.**

Before lighting the burner, ask your instructor to check your setup. If the setup is satisfactory you may turn on the water to the condenser. Make sure that water flows smoothly into the drain. If the water pressure is excessive, the hose may separate from the condenser.

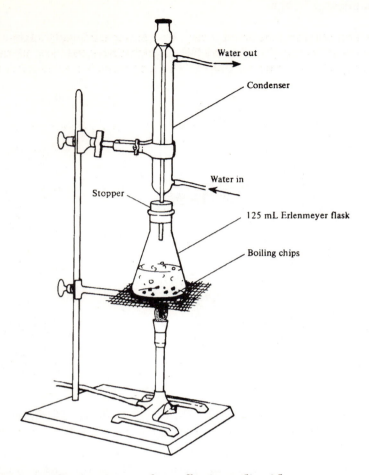

Figure 14. Apparatus for refluxing a liquid.

Light your burner. When the liquid in the flask starts to boil, decrease the gas supply to the burner so that the rate of boiling is slow to moderate. **Do not boil too rapidly.** A correct boiling rate is established when the top of the liquid inside the condenser is condensing about 5 cm above the water inlet. Monitor the apparatus during the reflux period. Make the adjustments necessary to maintain both a moderate boiling rate and a steady flow of water through the condenser.

G. USE OF THE CRUCIBLE

A crucible with its lid is a porcelain container used to run reactions at rather high temperatures. When used for quantitative work, the empty crucible is first heated to drive off residual moisture. Such moisture adds an unknown and variable weight to the empty crucible. After the crucible has cooled (in a desiccator), a reliable dry weight can be obtained. After running a reaction, the crucible and its contents are again heated to drive out moisture. After cooling, the crucible and its contents are weighed again.

Firing the Crucible

Support the crucible and the lid on a clay triangle (Figure 15). Heat the crucible in the hottest flame possible (read Section E, pages 11-12). The crucible must glow bright orange to assure adequate heating. Use the crucible tongs to adjust or move the hot crucible and lid.

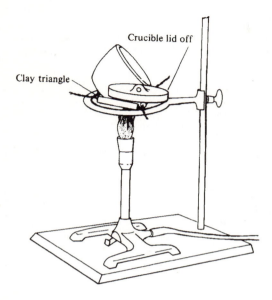

Figure 15. Drying and firing a crucible and cover.

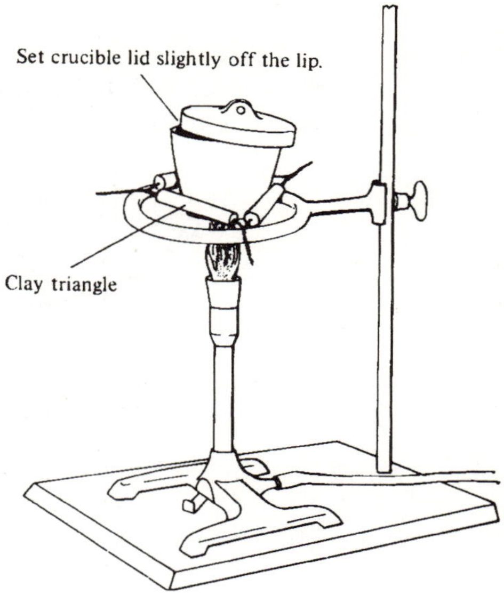

Figure 16. Ignition of the contents of a crucible.

Place a hot crucible for cooling on the ceramic mat (Figure 17). A hot crucible will either melt or char any other surface. After cooling, always use the tongs to move the crucible. Do not touch a fired crucible with your fingers. The oils from your skin will change the weight of the crucible. Use a casserole to hold the cool crucible and lid for transport to the balance.

Figure 17. Cooling hot objects.

Ignition of Crucible Contents

Set the crucible upright in the clay triangle with its lid slightly off (Figure 16). Follow the procedure given for firing the crucible in each experiment.

H. GLASSWORKING

You may need to construct experimental setups that require glass tubing. Unsafe techniques in glassworking cause many accidents. Before attempting the construction of any apparatus with glass tubing, study and review this section.

Cutting Glass Tubing

Place the piece of glass tubing to be cut on the bench. Use a glass scorer or the edge of a triangular file to make a scratch on the tube where it is to be broken (Figure 18). Do not saw the glass tubing. One single firm scratch is sufficient.

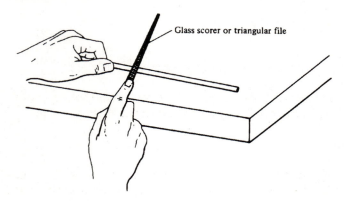

Glass scorer or triangular file

Figure 18. Scratching a piece of glass tubing.

Protect your hands by placing the tubing in a cloth towel. Grasp the tubing with your thumbs behind the scratch (Figure 19). While pulling slightly on each part of the tubing, push your thumbs gently outward. A clean break should result at the point of the scratch.

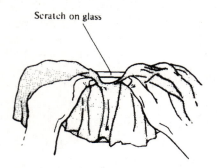

Scratch on glass

Figure 19. Breaking a glass tube.

Fire Polishing

The edges of freshly broken glass are very sharp. These sharp edges can cause severe cuts in your skin, rubber tubing, and rubber stoppers. The rough edges are made smooth by melting them (Figure 20).

Hold the cut end of the tubing in the hottest portion of the gas burner flame (Figure 11). Rotate the tube during heating to assure uniform melting. Remove the tube from the flame often and observe the end. Continue to heat the end of the tube until it is smooth. Avoid excessive fire polishing, since this causes a reduction in the diameter of the opening.

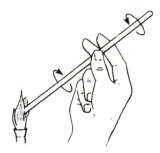

Figure 20. Fire polishing a glass tube.

Do not touch the hot end of the tube. Lay the tube on a ceramic mat until it has cooled. **Always use fire-polished tubing in constructing an experimental setup.**

Inserting Glass Tubing

Inserting glass tubing into a hole of a rubber stopper is frequently the cause of serious injury. Accidentally breaking the glass tubing can produce a sharp jagged edge. The broken glass then can slash through your hand during the attempted insertion. **A cloth towel must be used to protect both hands from this common source of serious injury.**

Moisten both the glass tubing and the hole in the stopper with glycerine or water. While protecting your hands with a cloth towel, place one hand on the glass tube 2-3 cm from the end (Figure 21). Place the other hand on the rubber stopper and simultaneously twist and push the tubing through the hole. Do this slowly and carefully to avoid breaking the glass. Do not rush. If any difficulty arises, consult your instructor. Rinse the tubing and stopper after insertion.

Figure 21. Inserting a glass tube into a stopper.

I. READING A MENISCUS

The exact measurement of the volume of a liquid in a graduated cylinder, a buret, or a pipet involves viewing the meniscus. The volume should be read at the bottom of the meniscus (Figure 22). The eye level must be horizontal to the surface of the liquid. Sighting either too high or too low will result in an inaccurate reading of the volume.

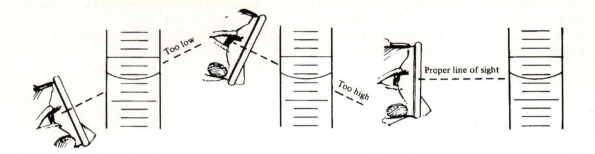

Figure 22. The eye must sight horizontally to the bottom of the meniscus.

J. VOLUMETRIC ANALYSIS

Volumetric analysis refers to a related group of procedures and equipment for measuring volumes precisely. With high quality glassware and careful techniques, volumes can be measured to a precision of a few parts per thousand. For example, a 10 mL volumetric pipet can deliver 10.00 mL repeatedly with an error of less than 0.02 mL. This error is less than ½ drop.

Titration is the cornerstone of volumetric analysis. The titration solution is added to an unknown from a buret. The unknown is titrated to determine the exact volume of the titration solution required to react completely with the unknown. After all of the unknown substance has reacted, the titration is stopped. The stopping point (endpoint) is signaled by a color change of an indicator. The exact concentration or quantity of the unknown is then calculated from the exact volume of the titration solution required to make the indicator change color.

Standard Solutions

The results of the calculation in a titration experiment depend on the concentration of the titration solution. To achieve an accurate result, the concentration of the titration solution must be known very accurately. A solution whose concentration is known very accurately is a standard solution. The error in the concentration of a standard solution is often only a few parts per thousand. The standard solutions used in your experiments have been carefully prepared.

Once a solution has been standardized, special precautions must be taken to avoid an accidental change in concentration. Do not leave the standard solution unstoppered. Do not return excess standard solution to the stock bottle. Do not dip pipets or any other item into the stock bottle. Always pour small amounts of the standard solution into a clean dry beaker for your use.

Any liquid unknown given to you for analysis by titration is actually a standard solution. This liquid unknown requires the same careful handling to avoid contamination.

Cleaning Volumetric Equipment

For accurate results both the pipet and buret must be clean. Clean glass is wetted uniformly by distilled water. Water forms drops on dirty glass. Your buret and pipet should be washed thoroughly with detergent solution, rinsed with tap water, and then rinsed thoroughly several times with distilled water. A special brush with a long handle is used to scour the inside of a buret. A pipet is cleaned with a pipe cleaner or pipet brush.

After cleaning, fill the pipet or buret with distilled water. Test the drainage. If droplets form inside the pipet or buret, repeat the cleaning procedure.

Using the Pipet

The volumetric pipet (Figure 23) is used to deliver an accurate volume of the unknown for titration. **Under no circumstances should you fill a pipet by sucking on it with your mouth.** A rubber bulb is used to draw liquid into a pipet, while the tip is placed sufficiently below the liquid in a beaker or a flask (Figure 23). Draw the liquid above the calibration mark, and then quickly remove the bulb and place your forefinger over the end. By rotating the pipet and releasing the pressure gently, the liquid is allowed to fall to the calibration mark. Touch the inside of the flask above the surface of the liquid to remove any drops on the tip.

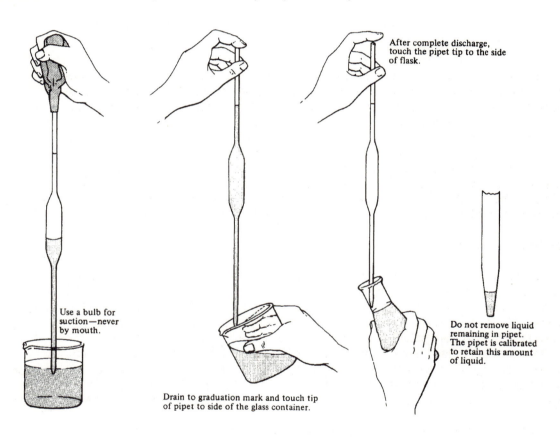

After complete discharge, touch the pipet tip to the side of flask.

Use a bulb for suction—never by mouth.

Drain to graduation mark and touch tip of pipet to side of the glass container.

Do not remove liquid remaining in pipet. The pipet is calibrated to retain this amount of liquid.

Figure 23. Use of the volumetric pipet.

Dispense the liquid into another flask by releasing your forefinger. Allow the liquid to flow freely. Touch the tip of the pipet to the inside of the flask to remove the last drop from the tip. Do not force out the slight amount retained within the tip of the pipet. You should practice using the pipet with distilled water.

Using the Buret

The buret is used to add a measured volume of a titration solution. Clean the buret thoroughly and fill it with a titration solution as shown in Figure 24.

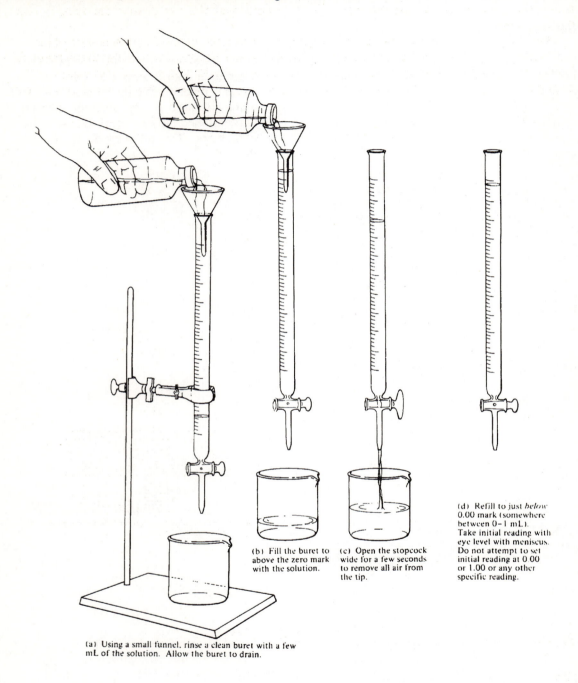

(a) Using a small funnel, rinse a clean buret with a few mL of the solution. Allow the buret to drain.

(b) Fill the buret to above the zero mark with the solution.

(c) Open the stopcock wide for a few seconds to remove all air from the tip.

(d) Refill to just *below* 0.00 mark (somewhere between 0–1 mL). Take initial reading with eye level with meniscus. Do not attempt to set initial reading at 0.00 or 1.00 or any other specific reading.

Figure 24. Filling a buret for a titration.

Filling. Place about 5 mL of the titration solution in the buret. Twist and turn the buret in a nearly horizontal position so that the solution covers all inner surfaces. Allow the rinse solution to drain through the stopcock and out the tip of the buret. Then fill the buret above the zero mark with fresh titration solution. Slowly open the stopcock and allow the solution to run into the tip. *Make sure there are no air bubbles in the tip.* Adjust the level of the solution in the buret to near the zero mark.

Reading. The volume of the liquid in the buret is measured by locating the bottom of the meniscus. The proper technique requires that the eye view exactly perpendicular to the buret by looking across a full circle mark on the buret (Figure 25). A buret reading card with a dark background is placed behind the buret to facilitate reading the scale. The example shown is read as 14.76 mL.

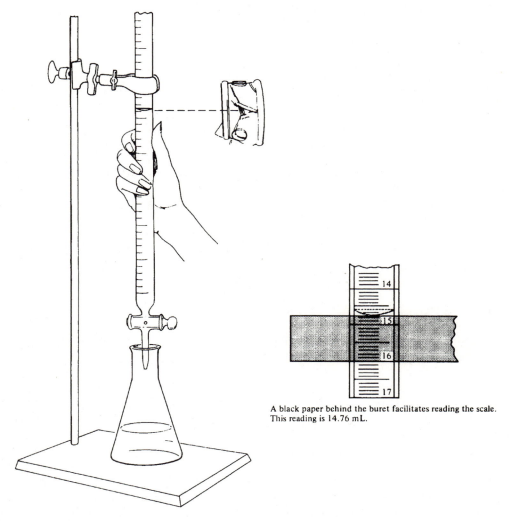

A black paper behind the buret facilitates reading the scale. This reading is 14.76 mL.

Figure 25. Reading the buret. Keep your eye level with the top of the liquid.

The volume dispensed in a titration is determined from the difference in the readings of the buret. Read the buret prior to the start of the titration and at the conclusion of the titration after the indicator has changed color.

Titrating. Place the tip of the buret slightly within the flask and control the addition of the solution as shown in Figure 26. Note the use of the left hand to control the stopcock.

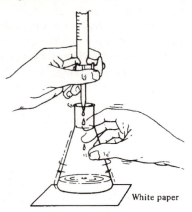

White paper

Figure 26. Titration procedure. Swirl the flask repeatedly during titration. A white paper under the flask accentuates the color change.

The flask should rest on a piece of white paper. Gently swirl the contents of the flask with your right hand. Do not shake the flask as liquid may splash out.

Use your wash bottle to wash the titration solution down the walls of the flask. Use small portions of 2-5 mL of distilled water . Washing is especially important near the endpoint. Carefully watch the contents of the flask to detect any color changes during the titration. As the color change becomes imminent, adjust the rate of addition of titration solution so that small drops are added one at a time.

K. MELTING POINT

The temperature at which a solid compound liquefies is its melting point. Since the melting point is an intrinsic property, it is useful in characterizing compounds. Many organic compounds have melting points in the range 50-300°C so that simple techniques for their determination can be employed. The melting points of inorganic compounds are not routinely taken because they are outside the range of simple techniques. Sodium chloride, for example, melts at 804°C.

For the organic chemist, determining the melting point is usually the first step in the identification of a substance. The melting-point technique can be used for an unambiguous identification of the unknown compound if a supply of known compounds is available. This approach is called the mixed-melting-point technique.

In a mixed melting point, one first determines the melting point of the unknown. Then known compounds that have a melting point within two or three degrees of the unknown are located. A small sample of each of the knowns is mixed with separate small portions of the unknown, and the melting points of these mixtures are determined. For a given mixture the melting point can be either the same value or lower than the melting point of the pure unknown.

When the mixed melting point has the same value as that of the pure unknown, the unknown and the known samples are identical and positive identification has been made. When any two different compounds are mixed together, the melting point of the mixture always is lower than the melting point of either pure compound. A mixture of two solid compounds may liquefy during mixing in preparation for a mixed-melting-point determination. The melting point of such a liquid mixture is below room temperature. This fact offers dramatic proof that the two compounds are not the same.

The melting point can also provide information about the purity of a compound. Since the melting point of a mixture is less than that of either pure substance, the melting point of an impure sample will be lowered by the presence of the impurity. When purifying compounds, chemists strive to obtain the highest possible melting point because it indicates the highest possible purity. In addition to having a lower melting point, an impure sample also has a wider melting-point range. The melting-point range is reported as the temperature at which melting commences followed by the temperature at which the last trace of solid has disappeared. For example, an impure sample may melt in the range 62-64°C, while after purification it may melt at 63.9-64.2°C.

Determination of the Melting Point

Only a small amount of a compound is required to determine its melting point. Use a spatula to crush a quantity about the size of a small grain of rice on a watch glass. Grind the particles small enough to fit into the melting-point capillary.

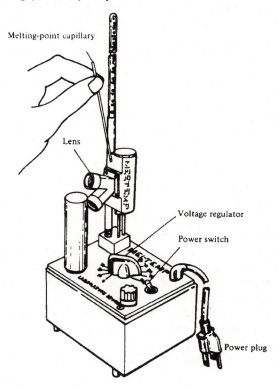

Figure 27. Melting-point apparatus.

Press the open end of a capillary tube into a portion of the compound to force the material into the tube. Invert the capillary tube and carefully tap it with your finger in order to vibrate the sample to the bottom of the tube. Repeat this procedure until the compound in the bottom of the tube reaches a depth of about 2 mm.

Turn on the power switch of the melting-point apparatus (Figure 27) and set the voltage to obtain the desired rate of heating for the anticipated melting range. Use a heating rate chart to select a voltage to obtain a heating rate of about 6 degrees per minute at the melting range.

Insert the sample in the slot near the thermometer. Observe the sample with your eye about 6 inches from the lens. Best results are obtained if the sample is melted slowly. If the voltage is set too high, a temperature lag will result and the recorded temperature will be lower than the true melting temperature.

Remove the capillary tube. Turn off the power switch; failure to do so can cause the instrument to overheat and cause damage.

L. SPECTROPHOTOMETRY

A colored solution absorbs light in the visible range of wavelengths. Consider a test tube containing a colored solution (Figure 28). The light of a specific wavelength is incident on the test tube from the left. The intensity of the light that leaves the tube on the right is diminished. The amount of light absorbed by the solution in the test tube is called the absorbance.

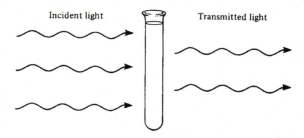

Incident light   Transmitted light

Figure 28. Absorption of light by a solution.

The absorbance of a solution is proportional to its concentration. A spectrophotometer is used to measure absorbance. Solutions of known concentrations are used to determine the proportionality of concentration and absorbance. The concentration of any other sample then can be calculated from its absorbance.

Use of the Spectrophotometer

The features of a spectrophotometer are shown in Figure 29. The instructions given below for the proper use of the spectrophotometer should be followed carefully. **Do not make any adjustments that you do not understand**. The instrument contains expensive electronic and optical parts that must be handled with care to prevent damage.

1. Close the sample cell holder cover.
2. Set the wavelength dial to the value desired.
3. Turn on the instrument and allow it to warm up for 10 minutes.

Figure 29. Spectronic 21. [Courtesy of Spectronic Instruments, Inc.]

4. Obtain a cuvette from your instructor.
5. Insert a reference cuvette, containing water (Figure 30), close the cell holder cover. When using the cuvette, always place it in the sample holder in the same orientation (etch mark toward the front). Always use the same cuvette for the reference cell and the sample cell.

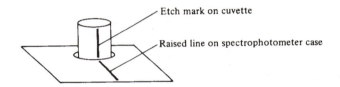

Figure 30. Correct insertion of the cuvette.

6. Set the light control to obtain 0 (zero) absorbance on the right side of the lower scale in Figure 31. Then remove the reference cuvette.
7. Insert each sample for the experiment, close the cell holder cover, and determine the absorbance of the sample.

Figure 31. The Spectrophotometer Scale.

M. USE OF THE LABORATORY BAROMETER

Atmospheric pressure is measured with a barometer. A barometer (Figure 32) is constructed by filling a long glass tube, closed at one end, with mercury. The open end of the filled tube is then inverted into a reservoir of mercury. The pressure of the atmosphere is exerted on the mercury surface in the reservoir. This pressure supports a length of mercury inside the tube. This length, measured from the surface of the reservoir to the top of the mercury meniscus inside the tube, is the atmospheric pressure.

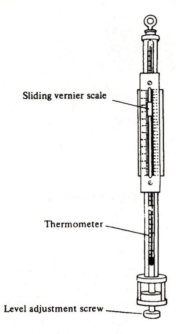

Sliding vernier scale

Thermometer

Level adjustment screw

Figure 32. A laboratory barometer.

To measure the atmospheric pressure in the laboratory, first establish the reference level in the reservoir. Adjust the level adjustment screw until the surface of the mercury in the reservoir just touches the reference pointer (Figure 33).

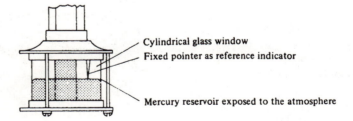

Cylindrical glass window
Fixed pointer as reference indicator

Mercury reservoir exposed to the atmosphere

Figure 33. Closeup of the mercury reservoir of a laboratory barometer.

Determine the height of the top of the mercury meniscus by aligning the sliding vernier scale with the meniscus (Figure 34). The atmospheric pressure then is read from the fixed and sliding vernier scales.

26

The barometer reading in Figure 34 is 74.83 cm Hg. The last place is given by the line on the sliding scale that is in best alignment with any line on the fixed scale.

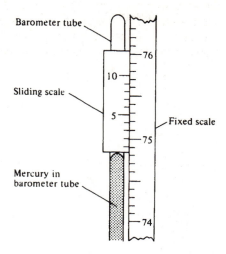

Figure 34. A barometer reading 74.83 cm of Hg.

N. USE OF THE BALANCES

Since many of your laboratory experiments depend upon your ability to make rapid and accurate measurements of mass on the balances, it is essential that you learn to use the instruments properly as soon as possible. You will use a top-loading balance accurate to 0.01 g and an analytical balance accurate to 0.1 mg. Your instructor will show you how to use the balances during the first laboratory period and you will then be asked to determine the mass of an object. If you experience difficulty in weighing the object, ask your instructor for additional help. The features of the top-loading balance are shown in Figure 35, those of the analytical balance in Figure 36.

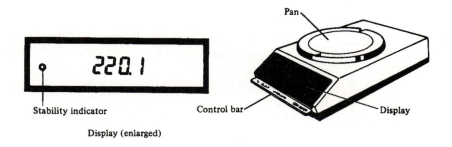

Figure 35. Top-loading balance.

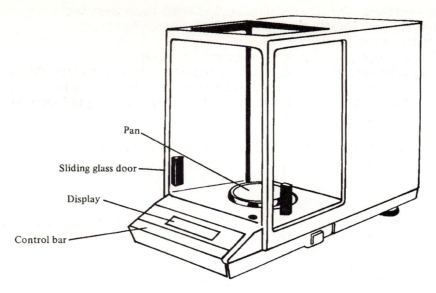

Pan

Sliding glass door

Display

Control bar

Figure 36. Analytical balance

Care of the Balances

Balances are expensive, precision-made instruments. They do not function properly if abused. Each student is responsible for cleanliness, knowledge of proper operating steps, and adherence to the following rules.

1. Never move the balance. Check to see that the balance is level, as indicated by the leveling bubble. If it is not level, ask the instructor for assistance.
2. Always close the sliding panels of the balance case. Air currents must be avoided.
3. Do not overload the balance. The maximum loads of the analytical and top-loading balances are about 100 and 600 g, respectively. (See your instructor for specific operating instructions for the balance that you are using.)
4. Never weigh any chemical or moist object directly on the balance pan. Immediately clean up any material that is spilled on the pans or within the balance.
5. Never place a hot object on the balance pan; its apparent weight will be incorrect because of convection currents set up by the rise of heated air.
6. Avoid using your fingers to handle objects to be weighed. Do not touch the balance pan or the pan support. Care must be taken to prevent weight changes caused by absorption of moisture or oil from the skin.
7. When weighing is completed, remove all objects from the pan, close the balance case and rezero the balance.
8. Check to see if you have accidentally spilled any material on or near the balance. If you have, clean up the spill before leaving.

Steps in Weighing

The weighing procedure consists of these steps: zeroing the balance to be sure the balance is properly adjusted; placing the object to be weighed on the pan; and—after waiting for the stability indicator to go out—reading the mass of the object from the display.

1. Checking the zero point. This must be done each time the balance is used.
    a. Unload the pan. Clean it if necessary, using a camel's-hair brush.
    b. Close the sliding panels of the balance case.
    c. Switch on the display by pressing the control bar once.
2. Placing the object on the pan. The object to be weighed is placed on the pan. Whenever possible, use a pair of tongs or tweezers to avoid transmitting moisture and oil from your skin to the object to be weighed. After introducing the object to be weighed, close the sliding panels of the balance immediately.
3. Reading the mass.
    a. Wait until the display is stable and the stability indicator goes out.
    b. Read the mass.

Taring and Weighing

It is not always necessary to determine the mass of a container in order to obtain the mass of a sample by difference. The direct determination of the mass of a sample after setting the mass of the container is called taring. The procedure is
1. Place the container on the pan.
2. Press the control bar and the balance will display zero.
3. Remove the container from the pan and add the sample.
4. Return the container to the pan and record the displayed mass.
5. Remove the container and sample and rezero the balance by pressing the control bar.

O. USE OF THE DIGITAL THERMOMETER

The digital thermometer is shown in Figure 37 on the following page.

1. Connect the thermocouple.
2. Turn on the thermometer. After a self-test display appears, the temperature is displayed in about 3 seconds.
3. Select the temperature scale by pressing the F/C button.
4. Pressing the hold key retains the last reading on the display. Pressing the key again causes the thermometer to resume measuring the temperature.
5. Conserve battery power by turning off the thermometer when it is not in use.

P. GRAPHING

Preparing a Graph

A properly drawn and labeled graph can communicate a great deal of information. The important features of a good graph include

1. The title.
2. The complete labeling of the x-axis or abscissa.
3. The complete labeling of the y-axis or ordinate.
4. The choice of appropriate scales for both the abscissa and the ordinate.
5. The experimental points or data points.
6. The drawing of a smooth, continuous curve (or straight line) through the points.

These features are illustrated in Figure 38 on page 31.

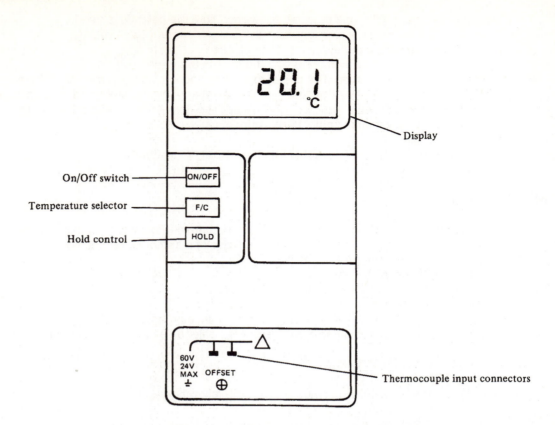

Figure 37. Digital thermometer.

The procedure for drawing an acceptable graph is as follows. First decide what is to be represented on the abscissa (x-axis) and on the ordinate (y-axis). Next choose a proper scale for each axis. Examine the maximum number and minimum number to be entered on each axis. Determine the value for each scale division *so that the data are spread across a major portion of the paper*. Do not squeeze the data points into a small corner. Do not allow the data to be spread along only one side of the graph. The choice of the scales for the axes is the critical step in drawing a graph.

After the scales have been determined, number the divisions along each axis. Also label each axis to avoid errors in entering the points. Now enter each experimental point.

Draw a circle around each experimental point. Connect the points by sketching a smooth curve that approaches all points as closely as possible. The curve need not pass exactly through every point. This usually is impossible because of experimental errors. Never connect the points with a series of straight lines that go through every point similar to a child's puzzle of connecting numbered dots. If the graph is supposed to be a straight line, use a transparent straight edge to draw one straight line that approaches all points equally (Figure 39). Finally, title the graph and be sure that each point and each label are entered correctly.

Several sheets of graph paper are printed at the back of this manual for use in the preparation of graphs for your laboratory reports.

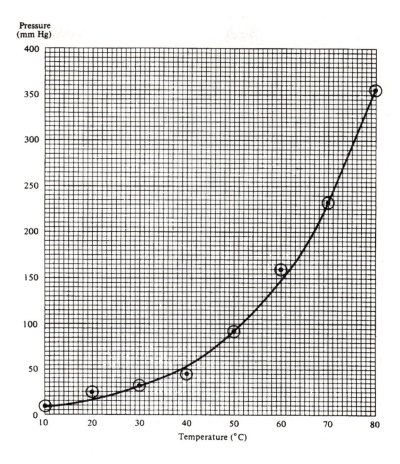

Figure 38. Vapor pressure versus temperature of water.

Calculation of the Slope of a Straight Line

To calculate the slope of a straight line, choose two points that are far apart (*not two data points*) and whose coordinates (x and y values) are easily read from the graph (see Figure 39 on the next page). Calculate the change in the x value, or Δx. Δx (pronounced "delta ex") is the difference in x values for these two points, $x_2 - x_1$. Calculate Δy, the difference in y values for the same chosen points, or $y_2 - y_1$. For the line of the graph in Figure 39, the slope is calculated as follows.

1. Read the x and y values for the chosen points from the graph.

    $x_2 = 6.0 \text{ cm}^3 \quad y_2 = 17.4 \text{ g}$
    $x_1 = 0.9 \text{ cm}^3 \quad y_1 = 2.6 \text{ g}$

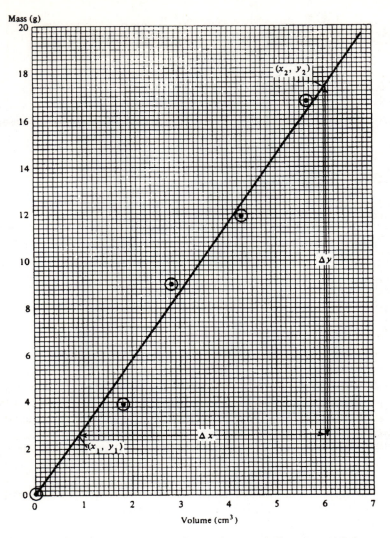

Figure 39. Mass versus volume for an unknown solid.

2. Calculate Δx.

$$\Delta x = x_2 - x_1 = 6.0 \text{ cm}^3 - 0.9 \text{ cm}^3 = 5.1 \text{ cm}^3$$

3. Calculate Δy.

$$\Delta y = y_2 - y_2 = 17.4 \text{ g} - 2.6 \text{ g} = 14.8 \text{ g}$$

4. Divide Δy by Δx to obtain the slope.

$$\text{Slope} = \frac{\Delta y}{\Delta x} = \frac{y_2 - y_1}{x_2 - x_1} = \frac{14.8 \text{ g}}{5.1 \text{ cm}^3} = 2.9 \text{ g/cm}^3$$

Significant Figures and Units

Exponential Numbers

Very large or very small numbers are encountered in scientific work. For example, light travels at a speed of roughly 30,000,000,000 cm/sec, and with an electron microscope a sphere with a diameter of 0.00000020 cm can be seen. Numbers such as these should be expressed as a coefficient times a power of 10. You will recall the following relationships.

$$1000 = 10^3 \qquad\qquad 0.1 = 10^{-1}$$

$$100 = 10^2 \qquad\qquad 0.01 = 10^{-2}$$

$$10 = 10^1 \qquad\qquad 0.001 = 10^{-3}$$

$$1 = 10^0 \qquad\qquad 0.0001 = 10^{-4}$$

This series could be extended in either direction. Since 400 is equal to 4×100, it can be written as 4×10^2. In the same way, 0.004 is equal to 4×0.001 and can be expressed as 4×10^{-3}. The numbers 4×10^2 and 4×10^{-3} are in exponential form. The number 4 in each case is the coefficient; the powers of 10 are the exponents.

Below is a review of some rules and examples of their use that will be helpful to you in working with exponential numbers.

1. In the exponential form of a number greater than 1, the power of 10 is positive and numerically equal to the number of places that the decimal point has been shifted to the left. Thus, 4300 is equal to 4.3×10^3, and the speed of light can be expressed as 3.0×10^{10} cm/sec.

2. In the exponential form of a number less than 1, the power of 10 is negative and numerically equal to the number of places that the decimal point has been moved to the right. Thus, 0.00025 is the same as 2.5×10^{-4}, and the particle size the electron microscope can resolve can be expressed as 2.0×10^{-7} cm.

3. To multiply exponential numbers, multiply the coefficients and add the exponents algebraically.

$$(2.0 \times 10^3)(4.0 \times 10^4) = (2.0 \times 4.0) \times 10^{(3+4)} = 8.0 \times 10^7$$

$$(3.0 \times 10^4)(2.0 \times 10^{-5}) = (3.0 \times 2.0) \times 10^{(4-5)} = 6.0 \times 10^{-1}$$

4. When dividing exponential numbers, divide the coefficients and algebraically subtract the exponent of the denominator from that of the numerator.

$$\frac{6.0 \times 10^5}{3.0 \times 10^2} = \frac{6.0}{3.0} \times 10^{(5-2)} = 2.0 \times 10^3$$

$$\frac{8.0 \times 10^3}{4.0 \times 10^{-7}} = \frac{8.0}{4.0} \times 10^{3-(-7)} = 2.0 \times 10^{10}$$

5. It is customary to express very large or very small numbers in exponential form with only one digit to the left of the decimal point. This is known as standard exponential form or standard scientific notation.

$$(3.0 \times 10^4)(4.0 \times 10^{-6}) = (3.0 \times 4.0) \times 10^{(4-6)} = 12 \times 10^{-2} = 1.2 \times 10^{-1}$$

Exact Numbers

Some quantities, such as the number of items in sample established by a direct count, are exact number. There is no uncertainty in the quantity. Other exact numbers are in defined equivalences such as 12 inches in 1 foot. Exact numbers do not have to be considered in determining the number of significant figures in a calculation in which they are used.

Significant Figures

Since all measurements represent a comparison to a standard, there is some degree of uncertainty in all measured quantities. For instance, suppose a crucible is weighed on a balance that is accurate to the nearest 0.01 g and that the mass of this crucible is determined to be 17.32 g. This means that the actual mass of the crucible is between 17.31 g and 17.33 g. There is some degree of doubt about the figure in the hundredths place. This fact may be represented when recording the mass as 17.32 ± 0.01 g. However, the ± 0.01 is usually not used. It is sufficient to recognize that the last recorded number of the quantity has an uncertainty of +1 or −1.

Now suppose that the same crucible is weighed on a balance accurate to the nearest 0.0001 g and that the mass of the crucible is determined to be 17.3247 g. This means that the mass is between 17.3246 g and 17.3248 g. The second measurement is more accurate than the first, but there is still some doubt about the last decimal place. The mass is know to ± 0.0001 g. Again, the 17.3247 g is recognized as having an uncertainty of +1 or −1 in the last recorded place.

It is important to indicate the accuracy with which a given quantity is measured. This is done by the number of significant figures in the measured quantity. The number 17.32 g contains four significant figures, whereas 17.3247 g contains six significant figures.

The number of significant figures expressed in a number are determined by the following rules.

1. Numbers that are determined by counting are exact numbers and can be assumed to have an infinite number of significant figures.

2. Nonzero integers in a measured quantity are always significant figures regardless of the position of the decimal point.

Zeros in Significant Figures

It is important to recognize whether a zero in a quantity is a significant figure or not. The zero may serve as a number or merely be included to locate the decimal point in a number. In the latter case, the zeros are not significant. The rules that allow you do determine whether a zero is significant or not are

1. Zeros between two nonzero digits, called captive zeros, are significant figures.

2. Zeros that precede all of the nonzero digits, called leading zeros, are not counted as significant figures.

3. Zeros to the right end of a number, called trailing zeros, are significant if the number contains a decimal point.

The quantities 3.48, 34.8, and 348 all contain three significant figures. Both 3.041 and 3.504 contain four significant figures. The zeros in 0.34, 0.048, and 0.0057 are not significant (rule 4), and all three quantities contain two significant figures.

The quantity 3400 would contain four significant figures if the precision of its measurement was in the units position. In this case, we could write $3400 = 3.400 \times 10^3$. Similarly, if it were known that the precision of 3400 was only ±10, then it would be written $3400 = 3.40 \times 10^3$, which has three significant figures. Finally if the precision of 3400 were ±100, then it would be written $3400 = 3.4 \times 10^3$, which has two significant figures.

For numbers larger than 10 having trailing zeros, the assignment of significant figures is not possible without additional information. If the previous example were written with a trailing decimal point (3400.), this would unambiguously indicate four significant figures.

In calculations with measured quantities, the answer may not be more accurate than the least accurately measured quantity involved. In calculations involving the multiplication and division of measured quantities, the number of significant figures in the answer must be the same as the number in the least accurately measured quantity. For instance, if an object is found to have a mass of 14.624 g and a volume of 7.2 cm^3, the density of this object cannot be calculated to more than two significant figures. The mass was determined to five significant figures, whereas the volume was determined to two significant figures; thus, the density can only be calculated to two significant figures. Using more than two significant figures would have no meaning because one of the quantities was not measured accurately enough to justify them. Accuracy is increased by more accurate measurement in the laboratory, not with pencil and paper.

In addition and subtraction operations, the last decimal place to be retained should correspond to the last digit in the least accurately measured quantity, as indicated by the underlined digits in the examples below.

$$
\begin{array}{ll}
2.3251\ \text{g} & \\
0.3\underline{3}\quad\ \text{g} & 4.7415\ \text{g} \\
0.616\ \ \text{g} & -2.\underline{2}\ \text{g} \\
\hline
3.27\quad \text{g} & 2.5\quad\ \text{g}
\end{array}
$$

Rounding Off Nonsignificant Figures

The process of rounding off numbers discards the parts of a number that are not significant. The rules for rounding off are

1. If the first nonsignificant figure is less than 5, drop it and all other nonsignificant figures.

2. If the first nonsignificant figure is more than 5 or is 5 followed by digits other than all zeros, drop all nonsignificant figures and increase the last significant figure by 1.

3. If the first nonsignificant figure is 5 alone or is 5 followed by only zeros, drop all nonsignificant figures and increase the last significant figure by 1 if it is odd but leave it alone if it is even.

Use of Units

Make it a habit when performing mathematical operations to label all quantities with the proper units and carry these units along in all calculations. Units can be multiplied, divided, squared—in fact, units can be made to undergo any operation that a number would undergo in algebra.

This use of units can be helpful in solving problems. Suppose that you know that the density of a substance is 0.927 g/cm^3 and wish to know what volume in cm^3 a 12.2 g sample of this substance will occupy. The setup for this calculation is (12.2 g)/(0.927 g/cm^3), which gives the answer in cm^3, the desired unit. A setup such as (0.927 g/cm^3) (12.2 g) can be seen to be wrong because the answer comes out in g^2/cm^3!

QUESTIONS

1. Express the following numbers in standard scientific notation.

   a. 470.9 _____   **c.** 5604 _____

   b. 0.0150 _____   **d.** 0.0031 _____

2. Express the following as ordinary numbers.

   a. 3.45×10^2 _____   **c.** 6.912×10^{-2} _____

   b. 4.1×10^{-3} _____   **d.** 6.62×10^1 _____

3. Suppose that each of the numbers in Questions 1 and 2 represented a measured quantity. How many significant figures would there be in each of these quantities?

   2a _____   **3a** _____

   2b _____   **3b** _____

   2c _____   **3c** _____

   2d _____   **3d** _____

4. Metallic gold sells for about $14 per gram. Assume that you are using an analytical balance to determine the mass of gold that you wish to sell. How many dollars error is produced by a balance error of 0.0001 g? How many dollars error is produced by the error of 0.02 g with the top-loading balance? Is the mass determination on the top-loading balance satisfactory for this purpose?

5. Solve the following problems, and express all answers to the correct number of significant figures. Be sure to include units in all operations!

    a. A helium atom weighs 6.65×10^{-24} g. How many helium atoms are contained in a 1.0 g sample?

    b. The speed of sound is 3.40×10^4 cm/sec. What distance (in meters) can sound travel during a 48-minute chemistry lecture?

    c. The density of platinum is 21.4 g/cm^3. What is the mass if a cube of platinum 2.0 cm on a side in grams? In kilograms?

5. A crucible has a mass of 24.563449 g. If weighed on a top loading balance (accurate to 0.01 g) what would be the mass? If weighed on an analytical balance (accurate to 0.0001 g) what would be the mass?

EXPERIMENT

1 Measurements in the Laboratory

APPARATUS

Crucible, crucible cover, 10 mL graduated cylinder, 10 mL pipet, eye dropper, analytical balance, top-loading balance

INTRODUCTION

In this experiment you will make measurements of mass using two types of balances and measurements of volume using two types of volumetric glassware. You will express measured quantities using the proper number of significant figures and evaluate the precision and accuracy of the measurements.

Precision

Every measurement has a degree of uncertainty associated with it that the scientist must strive to reduce to an acceptable level. The analysis is often repeated several times to obtain a replicate set of measurements for comparison. The reproducibility of a result obtained by the same experimental method is termed **precision**. If the variations between individual measurements are sufficiently small, then an average of the measurements called the mean is calculated. The mean is obtained by summing the individual measurements and dividing by the number of the measurements. For a series of measurements x_1, x_2, and x_3 the mean is x_{av} given by the following expression.

$$\frac{x_1 + x_2 + x_3}{3} = x_{av}$$

The difference between any individual measurement and the mean is used to express the precision of the experimental value. The deviation of a measurement from the average is the absolute value of $x_1 - x_{av}$. The relative deviation is the quotient of the deviation for a measured quantity and the mean multiplied by 100%.

$$\frac{x_1 - x_{av}}{x_{av}} \times 100\% = \text{relative deviation}$$

The relative deviation is one method of expressing the precision of a measured quantity.

Accuracy

The **accuracy** of a measurement reflects its nearness to the "true" value, x_{true}. The difference between a measurement x_1 and x_{true} is the error of the measurement.

$$x_1 - x_{true} = \text{error}$$

Note that accuracy is not the same thing as precision, which only compares one measured quantity with others. The "true" value is seldom known—that is the reason why we do experiments, namely to determine a value. Thus, it is difficult to arrive at an estimate of the error of a measurement. The relative error is a quotient of the difference between a measured quantity and the true value divided by the true value multiplied by 100%.

$$\frac{x_1 - x_{true}}{x_{true}} \times 100\% = \text{relative error}$$

The relative error is one method of expressing the accuracy of a measured quantity.

Classes of Errors

Errors in measured quantities are the result of two broad categories termed determinate errors and indeterminate errors. **Determinate errors** are systematic errors whose source is assignable, such as an instrumental error, a personal error, or an error in the experimental method. Determinate errors affect the accuracy of the measurement. Instruments have physical limitations such as where the markings on a scale are placed or how the instrument responds to electricity. Determinate errors then are usually all in one direction for a given physical measurement. Personal errors are often systematic as well. If one consistently sights a scale either too high or too low, then the errors will be in one direction. Finally, the experimental method selected may have a consistent error such as an incomplete reaction that gives only 99.5% of the possible product.

Indeterminate errors are random, which means that they cannot be easily identified. These types of errors are usually not in the same direction. One common source of indeterminate error is the variation in the technique of the investigator. Given a sufficient number of measurements, this type of error tends to average out, and the average value may approach the "true" value. The magnitude of this type of error is reflected in the precision of the measurement.

PROCEDURE

A comparison of mass measurements obtained with a top-loading balance to those obtained with an analytical balance will illustrate significant figures and accuracy. Record all mass data on the Report Sheet.

A comparison is made of the volumes delivered by graduated cylinder and a pipet by using the mass of the sample and the density to calculate the volume. Record all data on the Report Sheet.

Part A—Mass and the Top-Loading Balance

Study Section N, "Use of the Balances," on pages 27-29 of this manual. Then determine the mass of each of the following, using the top-loading balance.

(a) Crucible alone.
(b) Crucible lid alone.
(c) Crucible and crucible lid together.

For each measured quantity, indicate the number of significant figures. Compare the sum of (a) and (b) with the directly determined quantity (c). Are they the same? Account for any difference.
Determine the mass of each item (a), (b), and (c) again as a second trial. Compare the sum of (a) and (b) with the directly determined quantity (c). Are they the same? Account for any difference.
Account for any differences between the two trials of the masses of the crucible and lid. Calculate the average value. Divide the difference between each measured quantity and the average value by the average value and multiply by 100% to obtain the relative deviation of each quantity.

Part B—Mass and the Analytical Balance

Study the procedure and instructions for using the analytical balance in Section N (pages 27-29) of this manual. Repeat the mass determinations (a), (b), and (c) given in Part A using the analytical balance. Record all mass data on the Report Sheet. For each measured quantity, indicate the number of significant figures. Compare the sum of (a) and (b) with the directly determined quantity (c). Account for any difference.
Determine the mass of each item (a), (b), and (c) again as a second trial. Compare the sum of (a) and (b) with the directly determined quantity (c). Are they the same? Account for any difference.
Account for any differences between the two trials of the masses of the crucible and lid. Calculate the average value. Divide the difference between each measured quantity and the average value by the average value and multiply by 100% to obtain the relative deviation of each quantity.
Compare the results obtained using the analytical balance with those obtained using the top-loading balance. How do they differ?

Part C—Volume and the Graduated Cylinder

Study the procedure and instructions on page 21 of Section J concerning the location of the bottom of a meniscus and reading the volume of a liquid.
Add about 50 mL of distilled water to a beaker or Erlenmeyer flask and determine the temperature of the water. Record the temperature and the corresponding density of water given in the table on the following page.

Measurements in the Laboratory

Table 1.1

temperature (°C)	density (g/mL)	temperature (°C)	density (g/mL)
15	0.9991	21	0.9980
16	0.9990	22	0.9977
17	0.9988	23	0.9975
18	0.9986	24	0.9973
19	0.9984	25	0.9970
20	0.9982	26	0.9968

Using this sample, add distilled water to a 10.0 mL graduated cylinder to slightly below the 10.0 mL mark. Use a dropper to add the few drops necessary to bring the level to the 10.0 mL mark. If you initially add more than required, remove some of the liquid using the dropper until the volume reads 10.0 mL. Determine the mass of a 50 mL beaker using the analytical balance. Pour the contents of the graduated cylinder into the beaker and determine the mass of the beaker and the water. Using the density of water obtained from Table 1.1 at the measured temperature, calculate the volume of water delivered to the beaker. This is done by dividing the mass by the density.

Compare the calculated volume to the 10.0 mL that should have been delivered by the graduated cylinder. Divide the difference of these two quantities by 10.0 mL and multiply by 100% to calculate the relative error.

Discard the contents of the 50 ml beaker and dry it. Determine its mass. Repeat the experiment a second time. Compare the calculated volume to the 10.0 mL that should have been delivered by the graduated cylinder. Divide the difference of these two quantities by 10.0 mL and multiply by 100% to calculate the relative error. Compare the calculated volumes for the two trials. Calculate the average value. Divide the difference between each measured quantity and the average value by the average value and multiply by 100% to obtain the relative deviation of each quantity.

Part D—Volume and the Pipet

Study the procedure and instructions on page 19 of Section J concerning the use of the pipet. Determine the mass of a clean dry 50 mL beaker using the analytical balance. Using the same water sample as in Part C, pipet 10.00 mL of distilled water into the beaker. Touch the tip of the pipet to the side of the beaker to remove the last drop. (Note—do not force out the small amount retained within the tip of the pipet.) Determine the mass of the beaker and the water. Using the density of water obtained from the table of part C, calculate the volume of water delivered to the beaker. This is done by dividing the mass by the density.

Compare the calculated volume to the 10.00 mL that should have been delivered by the pipet. Divide the difference of these two quantities by 10.00 mL and multiply by 100% to calculate the relative error.

Discard the contents of the 50 ml beaker and dry it. Determine its mass. Repeat the experiment a second time. Compare the calculated volume to the 10.00 mL that should have been delivered by the pipet. Divide the difference of these two quantities by 10.00 mL and multiply by 100% to calculate the relative error. Compare the calculated volumes for the two trials. Calculate the average value. Divide the difference between each measured quantity and the average value by the average value and multiply by 100% to obtain the relative deviation of each quantity.

Report Sheet

Name	Instructor/Section	Date	

Part A—The Top-Loading Balance

	I	Significant Figures	II	Significant Figures
Mass of Crucible (a)	(1) ————	————	(1) ————	————
Mass of Crucible Lid (b)	(2) ————	————	(2) ————	————
Sum of (a) and (b)	(3) ————	————	(3) ————	————
Mass of Crucible and Lid	(4) ————	————	(4) ————	————
Difference between (3) and (4)	(5) ————		(5) ————	
Average Value of (4)	(6) ————			
Relative Deviation of (4)	(7) ————		(7) ————	

Part B—The Analytical Balance

	I	Significant Figures	II	Significant Figures
Mass of Crucible (a)	(1) ————	————	(1) ————	————
Mass of Crucible Lid (b)	(2) ————	————	(2) ————	————
Sum of (a) and (b)	(3) ————	————	(3) ————	————
Mass of Crucible and Lid	(4) ————	————	(4) ————	————
Difference between (3) and (4)	(5) ————		(5) ————	
Average Value of (4)	(6) ————			
Relative Deviation of (4)	(7) ————		(7) ————	

Measurements in the Laboratory

Part C—Volume and the Graduated Cylinder

	I	II
Mass of Beaker	(1) ——————	(1) ——————
Mass of Beaker and Water	(2) ——————	(2) ——————
Mass of Water	(3) ——————	(3) ——————
Calculated Volume of Water	(4) ——————	(4) ——————
Difference between 10.0 mL and (4)	(5) ——————	(5) ——————
Relative Error of (4)	(6) ——————	(6) ——————
Average Value of (4)	(7) ——————	
Relative Deviation of (4)	(8) ——————	(8) ——————

Part D—Volume and the Pipet

	I	II
Mass of Beaker	(1) ——————	(1) ——————
Mass of Beaker and Water	(2) ——————	(2) ——————
Mass of Water	(3) ——————	(3) ——————
Calculated Volume of Water	(4) ——————	(4) ——————
Difference between 10.00 mL and (4)	(5) ——————	(5) ——————
Relative Error of (4)	(6) ——————	(6) ——————
Average Value of (4)	(7) ——————	
Relative Deviation of (4)	(8) ——————	(8) ——————

Answer the questions on the next page.

QUESTIONS

1. Round off the value of the mass of the crucible and lid obtained by the use of the analytical balance and compare it with the value determined using the top loading balance. Discuss any difference observed.

2. What would be the effect in this experiment of touching the crucible several times during the mass determinations?

3. Consider the drainage characteristics of the pipet used in this experiment. Would any error introduced be determinate or indeterminate?

4. Identify an indeterminate error associated with the use of the graduated cylinder.

5. If the residue in the tip of the pipet were forced out, how would the mass of the "delivered" sample be affected?

6. If the internal diameter of the graduated cylinder is larger than it should be and the marking were maintained at the same distances as in a "correct" graduated cylinder, what would be the effect on the volume delivered?

7. If the graduated cylinder is not clean and a drop adheres to the inside surface, what would be the effect on the volume delivered?

8. Assume that you sight below the required parallel position when reading the graduated cylinder. What would be the difference between 10.0 mL and the actual volume delivered?

9. Assume that you also sight below the required parallel position when filling the pipet to the mark. Would the magnitude of the error be smaller or greater than that for the graduated cylinder?

EXPERIMENT

2   **Density**

APPARATUS

100 mL graduated cylinder, 10 mL graduated cylinder, analytical balance, medicine dropper.

REAGENTS

Four different-sized samples of the same solid substance. Choices of substances may be rubber stoppers, glass marbles, plastic chunks, or various types of metal chunks. Aqueous sodium chloride solutions of the following concentrations (weight percent): 4, 8, 12, and 16%. Various unknown sodium chloride solutions with concentrations between 4 and 16%.

INTRODUCTION

The mass of a sample of a solid substance is an extrinsic property of the solid; its value depends on the size of the sample studied. An analytical balance is used to determine the mass of a substance in grams.

The volume of a specific sample of a solid substance is also an extrinsic property. The volume of the substance may be determined by liquid displacement. A solid that is heavier than water will sink and displace a volume of liquid equal to its volume.

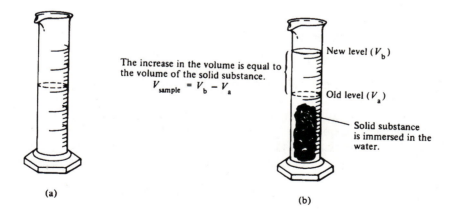

The increase in the volume is equal to the volume of the solid substance.
$$V_{sample} = V_b - V_a$$

New level (V_b)

Old level (V_a)

Solid substance is immersed in the water.

(a)   (b)

Figure 2-1. The increase in the volume is equal to the volume of the solid substance.

Although the standard unit of volume is the liter, the milliliter (mL) is generally used to express volume in the experiments in this manual. Volumes of solids are often expressed in cubic centimeters (cm^3). One cubic centimeter is equivalent to 1 mL.

The density of a substance is the ratio of the mass to the volume. Since the ratio does not depend on the size of the sample, density is an intrinsic property. In Part A of this experiment, the density of a solid substance is determined from a graph of mass against volume.

Whenever a straight line can be drawn through a set of data points, the data can be described by the mathematical expression $y = mx + b$. In this expression y represents the values plotted on the ordinate (vertical axis); x represents the values plotted along the abscissa (horizontal axis); m is the slope of the straight line; and b is the intercept of the straight line at the ordinate (the value of y when x is zero).

The relationship between the two properties of mass and volume for various samples of the same substance can be determined by a graphical approach. If the mass and the volume of each of the samples are determined, the mass of each sample can be plotted on the ordinate and the corresponding volume on the abscissa. The density of the substance can then be determined from this graph. Before preparing your graphs, study Section P, "Graphing," on pages 30-33 of this manual. Graph paper is provided at the back of this manual.

In Part B of this experiment, the densities of several solutions are determined from their mass and volume. The density can be related graphically to other properties. For example, a graph of the density of a solution versus the percent composition of the solute in the solution can be prepared. Such a graph will permit the indirect determination of the percent composition of an unknown solution by measuring its density.

PROCEDURE (Part A)

Obtain from your instructor several samples of the same solid substance and determine the mass of each sample. Record the masses in the table provided on the Report Sheet. No two samples can be of the same size. If the only sample pieces available are all the same size, "different" sizes may consist of

1. Piece 1 alone.

2. Pieces 1 and 2 together.

3. Pieces 1, 2, and 3 together.

4. Pieces 1, 2, 3, and 4 together.

Add enough distilled water to a clean 100 mL graduated cylinder so that the solid substance will be completely submerged after it is placed in the cylinder. Read and record on the Report Sheet the volume of the water to the nearest 0.1 mL. Be sure to read the bottom of the meniscus in determining the volume. (Study Section I, "Reading a Meniscus," on pages 17-18 of this manual.) Tilt the cylinder and gently slide the sample down the inside wall. Read, and record on the Report Sheet, the volume of the solid sample plus the volume of the water in the graduated cylinder. Calculate the volume of the solid sample by difference. Repeat the measurement for each sample.

Dry the samples and return them to the properly labeled container or to your instructor.

TREATMENT OF DATA (Part A)

Using the data entered in the Report Sheet, prepare a rough graph by plotting the mass of each sample on the ordinate versus its volume on the abscissa. Graph paper is provided at the back of this manual. Have your instructor check your graph prior to leaving the laboratory.

PROCEDURE (Part B)

Before performing this procedure, study Section A, "Handling Liquid Reagents," on pages 6-7 of this manual.

Obtain approximately 15 mL each of an unknown and four known sodium chloride solutions and place each in a labeled clean dry beaker or test tube. Each solution should be at room temperature. Note and record the temperature.

Thoroughly clean and dry a 10 mL graduated cylinder and determine its mass with the analytical balance. Transfer slightly less than 10 mL of the 4% sodium chloride solution to the 10 mL graduated cylinder, being careful not to spill any on the cylinder exterior. Using a medicine dropper, add sufficient 4% sodium chloride solution to bring the bottom of the meniscus to the 10.0 mL mark.

Determine the mass of the filled graduated cylinder on an analytical balance. Empty the graduated cylinder and rinse it with approximately 5 mL of the 8% sodium chloride solution. Discard the rinse, and proceed as described above for the 8% sodium chloride solution. Repeat the procedure for the remaining known solutions and the unknown. Do not use the taring feature of the balance when determining the mass of the cylinder containing 8%, 12%, 16%, and unknown sodium chloride solutions. The mass of each solution should be calculated by difference of the mass of the cylinder containing the solution and of the mass of the dry cylinder determined at the start of the procedure. Record all measurements on the Report Sheet.

TREATMENT OF DATA (Part B)

Using the data recorded on the Report Sheet, calculate the density of each of the four known solutions and prepare a graph by plotting density on the y-axis and the percentage of sodium chloride on the x-axis. Draw the best line through the points, and use the graph to determine the concentration of the unknown sodium chloride solution. Determine the intercept of the line.

Report Sheet

| Name | Instructor/Section | Date |

Part A

Description of the solid _____

	Sample number			
	1	**2**	**3**	**4**
Mass (g)	_____	_____	_____	_____
Total volume (mL)	_____	_____	_____	_____
Volume of water (mL)	_____	_____	_____	_____
Volume of sample (cm³)	_____	_____	_____	_____
Density of solid (from graph)	_____			

Part B

Code letter or number of unknown solution _____

Mass of 10 mL graduated cylinder _____   Temperature _____

	Solution number				
	1	**2**	**3**	**4**	**Unknown**
Concentration (weight %)	_____	_____	_____	_____	
Mass of graduated cylinder plus solution (g)	_____	_____	_____	_____	_____
Mass of 10 mL of solution (g)	_____	_____	_____	_____	_____
Density of solution (g/mL)	_____	_____	_____	_____	_____

Concentration of unknown (from graph)   _____

Attach sheets with calculations, the graphs for Parts A and B, and the pages that follow.

51

QUESTIONS

1. Write a mathematical equation describing the relationship between the mass and volume of a solid substance of Part A. Identify all symbols and use the correct units in your equation. Using your data from Part A, determine the value of the slope m, labeling your answer with the proper units. What relationship exists between the slope and the density?

2. What is your value of the intercept b in Part A of this experiment? What should be the value of b for a solid?

3. Which experimental measurement limits the accuracy of the calculated density in Part A?

4. Assuming the solid will fit into each of the cylinders, would a 10 mL graduated cylinder provide more or less accurate volumes than a 100 mL graduated cylinder? Explain.

5. Which individual sample used in Part A would provide the most accurate density if only a single mass and volume were determined?

6. Describe the limitations of determining the density of a solid in Part A if a liquid with density 1.8 g/mL were used instead of water.

7. Suppose that a single sample is used to determine the density of a given substance. If the mass of the sample of 11.97 ± 0.01 g and its volume is 4.9 ± 0.2 cm³, within what range of values would the density lie?

8. From the point of view of significant figures, discuss the advisability of using a top-loading balance for Part B. Assume that the top-loading balance is accurate to ± 0.02 g.

9. If 25.0 mL of the salt solution were used in a 25.0 mL graduated cylinder for each density determination in Part B, how would the accuracy of the density data be affected ?

10. Suppose your 10 mL graduated cylinder was inaccurate and actually contained 10.5 mL when it read 10.0 mL. How would this affect the determination of the percent composition of salt in the unknown?

11. Why was it suggested in Part B that the graduated cylinder be rinsed with the next solution to be used rather than with distilled water?

12. What is the value of the intercept *b* in Part B of this experiment? What value would be expected for *b* ?

EXPERIMENT
3

Specific Heat of a Metal

APPARATUS

8 oz styrofoam cup with lid, test tube (25 × 200 mm), gas burner, ring stand, clamp, 600 mL beaker, test tube holder, thermometer or a digital thermometer.

REAGENTS

Solid unknown metal.

INTRODUCTION

Heat capacity is the amount of heat required to raise the temperature of a sample by one Celsius degree. Since the heat capacity depends on both the size and nature of the sample, heat capacities of substances are reported per gram of substance. This is called the specific heat for that substance.

In 1819 Dulong and Petit observed an approximate relationship between the specific heat of solid metallic elements and their atomic weights. They recognized that the specific heat near room temperature of most solid metallic elements had the same value, $6.3 \text{ cal mole}^{-1} \text{ C}^{-1}$ ($1.05 \times 10^{-23} \text{ cal atom}^{-1} \text{ C}^{-1}$). This observation allowed early investigators to determine approximate atomic weights from heat capacity data.

In this experiment the heat capacity of a metal is determined by measuring the equilibrium temperature attained when the heated metal is added to cool water in a calorimeter. Then the atomic weight of the metal is calculated using the Law of Dulong and Petit.

PROCEDURE

Before performing this experiment, study the following sections of this manual: E, "The Gas Burner" (pages 11-12); F, "Heating Liquids" (pages 12-14); and N, "Use of the Balances" (pages 27-29).

Assemble the calorimeter as shown in Figure 3-1. Use an 8 oz styrofoam cup with a lid. The lid must be modified by punching a hole for the thermometer or digital thermometer. (The special thermometer is expensive and should be clamped to a ring stand to prevent breakage.)

Obtain a sample of unknown metal and weigh it to the nearest 0.01 g. Your metal sample should weigh about 30 g. Gently slide the metal into a test tube held at a 30° angle. Place the tube containing the metal in a beaker of water. The water level in the beaker should be several centimeters above the top of the metal contained in the test tube. Boil the water for 15 minutes to assure that the temperature of the metal is that of the boiling water.

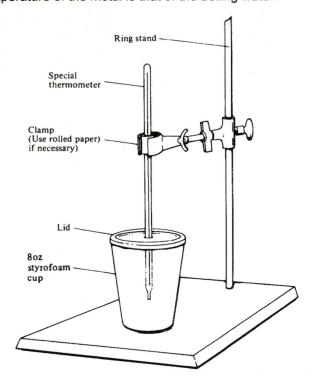

Figure 3-1. Calorimeter.

Cold

While the water is boiling, weigh the styrofoam cup (without lid) to the nearest 0.01 g. This cup will serve as a calorimeter. Place about 30 mL of water in the calorimeter and determine the weight of the calorimeter and water. Assemble the calorimeter and measure the temperature of the water to the nearest 0.1°C.

Remove the lid from the calorimeter. Remove the test tube from the boiling water and quickly slide the hot metal into the water in the calorimeter. Be careful not to allow any hot water clinging to the side of the test tube to enter. Replace the lid on the calorimeter and stir the water. Record the maximum temperature to the nearest 0.1°C. Record all data as Trial I.

Repeat the experiment using the same metal sample but place 35–40 mL of water in the calorimeter. Record all data as Trial II.

CALCULATIONS

When a hot metal is added to cool water, the water becomes hotter and the metal cools. The final temperature is between the initial temperature of the metal and the initial temperature of the water. If we assume the styrofoam cup is a perfect insulator and no heat energy leaves the system, then the loss of heat energy of the metal is equal to the gain of heat energy for the water. This idea can be expressed mathematically as heat energy lost by metal = heat energy gained by water

$$m_1 c_1 (\Delta T_1) = m_2 c_2 (\Delta T_2)$$

The subscripts 1 and 2 refer to the metal and water respectively; m is the mass in grams, c is the heat capacity, and ΔT is the change in temperature. For the metal, ΔT_1 is the temperature decrease, that is, the difference between the temperature of the hot metal and the final equilibrium temperature of the system. For the water, ΔT_2 is the temperature increase, that is, the difference between the temperature of the cool water and the final equilibrium temperature.

Since the specific heat of water (c_2) is 1.00 cal deg^{-1} g^{-1}, all of the quantities in the equation are known except c_1, the specific heat for the metal. From the calculated value for c_1, the atomic weight of the metal can be calculated using the law of Dulong and Petit.

$$c_1 \times \text{at. wt.} = 1.05 \times 10^{-23} \text{ cal atom}^{-1} \text{ C}^{-1}$$

These calculations are illustrated in the following example. Suppose 54.618 g of cadmium at an initial temperature of 100.0°C had been placed into a calorimeter that contained 40.498 g of water at 25.8°C. The temperature of the water in the calorimeter then increased from 25.8 °C to 30.9°C. The heat capacity of cadmium is calculated as follows.

$$
\begin{aligned}
\text{heat energy gained by water} &= m_2 c_2 \Delta T_2 \\
&= 40.498 \text{ g} \times (1.00 \text{ cal } °C^{-1} g^{-1}) \times (30.9 - 25.8)°C \\
&= 206.5 \text{ cal}
\end{aligned}
$$

heat energy lost by metal = heat energy gained by water = 206.5 cal
heat energy lost by metal = $m_1 c_1 \Delta T_1$

$$206.5 \text{ cal} = 54.618 \text{ g} \times c_1 \times (100.0 - 30.9)°C$$

$$c_1 = \frac{206.5 \text{ cal}}{54.618 \text{ g} \times (69.1°C)}$$

$$c_1 = 0.05471 \text{ cal } °C^{-1} g^{-1} = 0.055 \text{ cal } °C^{-1} g^{-1}$$
(to two significant figures)

The approximate atomic weight of cadmium then is calculated from the law of Dulong and Petit.

$$c_1 \times \text{at. wt.} = 1.05 \times 10^{-23} \text{ cal atom}^{-1} \text{ C}^{-1}$$

$$0.05471 \text{ cal } °C^{-1} g^{-1} \times \text{at. wt.} = 1.05 \times 10^{-23} \text{ cal atom}^{-1}$$

$$\text{at. wt.} = \frac{1.05 \times 10^{-23} \text{ cal atom}^{-1} \text{ C}^{-1}}{0.05471 \text{ cal } °C^{-1} g^{-1}}$$

$$\text{at. wt.} = 1.9192 \times 10^{-22} \text{ g atom}^{-1}$$

Specific Heat of a Metal

Convert the atomic weight expressed in units of g atom^{-1} to units of amu atom^{-1} using the defined quantity 1 amu = 1.6605×10^{-24} g.

$$\text{atomic weight} = (1.9192 \times 10^{-22} \text{ g atom}^{-1}) \times (1 \text{ amu}/1.6605 \times 10^{-24} \text{ g})$$

$$\text{atomic weight} = 115.58 \text{ amu/atom} = 1.2 \times 10^2 \text{ amu/atom}$$

(to two significant figures)

You may be puzzled by the use of 0.05471 cal °C^{-1} g^{-1} for the calculation of the atomic weight when only two significant figures are used in 0.055 for the specific heat. This illustrates the concern many chemists have with the propagation of errors. During a calculation that requires several steps, the intermediate values are not rounded off; only the final answer is rounded. This method prevents the accumulation of errors that may occur with premature rounding. In this example, the specific heat is properly reported as 0.055 cal °C^{-1} g^{-1} because ΔT_2 is known to only two signficant figures. The unrounded value, 0.05471 cal °C^{-1} g^{-1} is used for the further calculation of the atomic weight. However the calculated value for the atomic weight is then rounded to two significant figures.

Report Sheet

	Name	Instructor/Section	Date

	Trial I	Trial II

Mass of metal (g)   (1) _____   _____

Mass of water and calorimeter (g)   (2) _____   _____

Mass of empty calorimeter (g)   (3) _____   _____

Mass of water (g)   (4) _____   _____

Initial temperature water in calorimeter (°C)   (5) _____   _____

Initial temperature of hot metal (°C)   (6) _____   _____

Final temperature of water in calorimeter (°C)   (7) _____   _____

Final temperature of metal in calorimeter (°C)   (8) _____   _____

Change in temperature of water, ΔT_2 (°C)   (9) _____   _____

Change in temperature of metal, ΔT_1 (°C)   (10) _____   _____

Heat energy gained by water (cal)   (11) _____   _____

Heat energy lost by metal (cal)   (12) _____   _____

Specific heat of metal, c_1 (cal °C^{-1} g^{-1})   (13) _____   _____

Approximate atomic weight of metal (g/atom)   (14) _____   _____

Approximate atomic weight of metal (amu/atom)   (15) _____   _____

Average atomic weight of metal (amu/atom)   _____

Attach sheets with your calculations. Be sure that all of your data and calculations have the correct units and number of significant figures. Answer the questions on the next page.

Specific Heat of a Metal

QUESTIONS

1. The accuracy of what measurement must be increased in order to increase the accuracy for the specific heat? Explain why you made this choice.

2. Which metal, Pd or Pt, has the lower specific heat? Explain why, using the law of Dulong and Petit.

3. What would be the effect on your value of the specific heat if you did not leave the metal sample in the test tube long enough to heat the metal to 100°C?

4. How would the accuracy of your specific heat be affected if 200 mL of water were used rather than the suggested amounts?

5. How would the accuracy of your specific heat be affected if you used half the amount of the metal?

EXPERIMENT

4

Separation of a Mixture

APPARATUS

Two 100 mL beakers, stirring rod, funnel, ring stand, iron ring, clay triangle, filter paper.

REAGENTS

Acetone, solid unknown mixtures containing various known percentages of potassium chloride (KCl) and nickel carbonate ($NiCO_3$).

INTRODUCTION

For this experiment you will determine the percentage composition of a two-component solid mixture. Your unknown mixture will contain a water-soluble substance and a substance that is insoluble in water. An example is a mixture of potassium chloride and nickel carbonate. Potassium chloride is soluble in water, whereas nickel carbonate is not. This difference in solubility allows the components to be separated and the composition of the mixture to be determined.

The weight of the dried nickel carbonate after separation together with the original weight of the mixture allows the calculation of the weight of potassium chloride by difference. The actual amount of potassium chloride present in the water solution could be weighed after evaporation of the water, but this procedure is not included in this experiment.

PROCEDURE

Before performing this experiment, study the following sections of this manual: B, "Handling Solid Reagents" (pages 7-8); D, "Filtration" (pages 9-11); and N, "Use of the Balances" (pages 27-29). Obtain a mixture of unknown composition from your instructor. Determine to ±0.01 g the mass of a labeled 100 mL beaker. Transfer about 4 g of your unknown mixture into the beaker and determine the mass of the beaker and the sample (to ±0.01 g). Record all weights on the Report Sheet.

Prepare an apparatus for gravity filtration as shown in Figure 6 on page 9. Fold a piece of filter paper as shown in Figure 7 on page 9. Before wetting the paper, determine its mass to ±0.01 g. Be sure the mass is determined after the corner of the filter paper has been removed. Assemble the filter paper and funnel.

Add about 20 mL of distilled water to the beaker containing the solid unknown. Carefully swirl the slurry for about 2 minutes. Allow the solid to settle and then transfer the major portion of the liquid into the filter paper as shown in Figure 6 on page 9. Swirl the solid with two additional 20 mL portions of distilled water and filter the liquid. The extraction will proceed faster if the bulk of the solid is kept in the beaker. Only after several washings should the solid be transferred into the filter paper. Finally use small portions of water to wash all the solid remaining in the beaker into the filter paper. Allow all of the water to drain from the funnel.

Wash the solid and filter paper twice with 8 mL portions of acetone. The acetone dissolves the water and leaves mostly acetone in the filter paper. Acetone evaporates more rapidly than water, allowing quicker drying of the filter paper and the solid remaining in it. After the acetone has drained, carefully transfer and unfold the filter paper onto a clean dry watch glass. Cover the solid with a second sheet of filter paper and allow the solid to dry for at least 30 minutes. Placement of the sample into a hood will accelerate the drying process.

Weigh the dry filter paper and solid. Record the mass on the Report Sheet. Calculate the mass of the solid and determine the percentages of both components in the original mixture.

CALCULATIONS

Consider the following example of the method of calculation used in this experiment. Suppose a mixture of potassium chloride (KCl) and nickel carbonate ($NiCO_3$) weighed 3.57 g and, after treatment with water, 2.06 g of dry nickel carbonate remained.

$$\% \text{ component } = \frac{\text{mass of component}}{\text{total mass of mixture}} \times 100\%$$

$$\% \text{ NiCO}_3 = \frac{2.06 \text{ g}}{3.57 \text{ g}} \times 100\% = 57.7\%$$

$$\% \text{ KCl } = \frac{3.57 \text{ g} - 2.06 \text{ g}}{3.57 \text{g}} \times 100\% = 42.3\%$$

Report Sheet

_____ _____ _____
Name   Instructor/Section   Date

Mass of unknown solid and beaker (g)   (1) _____

Mass of empty beaker (g)   (2) _____

Mass of unknown (g)   (3) _____

Mass of insoluble solid and filter paper (g)   (4) _____

Mass of filter paper (g)   (5) _____

Mass of insoluble component (g)   (6) _____

Mass of soluble component (g)   (7) _____

% insoluble component   (8) _____

% soluble component   (9) _____

Attach sheets with your calculations. Answer the questions on the next page.

Separation of a Mixture

QUESTIONS

1. Explain how you would separate a solid mixture of table sugar from sand (SiO_2).

2. How could the technique used in this experiment be modified to separate a mixture of sand (SiO_2) and moth crystals (naphthalene). Neither sand nor moth crystals are soluble in water. Explain your proposed modification.

3. A student saved all the wash water, which contained all of the potassium chloride (KCl) from the unknown. Upon treatment of the wash water with excess silver nitrate ($AgNO_3$), a precipitate of silver chloride (AgCl) was formed. The AgCl was carefully collected, filtered, washed, dried, and weighed. If the weight of the AgCl is 3.97 g, what was the weight of KCl in the original sample?

EXPERIMENT
5   Analysis of Cations

APPARATUS

Test tube rack, 16 × 150 mm test tubes, medicine droppers, gummed labels.

REAGENTS

0.1 M solutions in dropping bottles of manganese(II) nitrate [$Mn(NO_3)_2$], copper(II) nitrate [$Cu(NO_3)_2$], iron(III) nitrate [$Fe(NO_3)_3$], cobalt(II) nitrate [$Co(NO_3)_2$], zinc nitrate [$Zn(NO_3)_2$], magnesium nitrate [$Mg(NO_3)_2$], and barium nitrate [$Ba(NO_3)_2$]. Solutions in dropping bottles of 2 M ammonium sulfate [$(NH_4)_2SO_4$], 2 M ammonium thiocyanate (NH_4SCN), 1 M ammonium sulfide [$(NH_4)_2S$], and 4 M sodium hydroxide ($NaOH$). Unknown solutions.

INTRODUCTION

Solutions of dissolved ionic materials may be analyzed to determine the presence or absence of specific cations. An analysis done to establish only the identity of the ions is qualitative analysis. An analysis to determine the amounts of each ion is quantitative analysis. In this experiment you will do a qualitative analysis of a cation in an unknown.

Qualitative analysis is based on using chemical tests that will give different results for each of several possible unknown materials. Ideally if a single reagent would react with only a specific cation, the observation of the expected reaction would unambiguously identify the unknown. The absence of a reaction would mean the absence of that specific cation. However, few reagents give specific tests for one cation of all the possible ions.

Qualitative analysis is then based on determining the response of an ion to several reagents. No two ions behave in exactly the same way when exposed to a variety of reagents. For example cations A^+ and B^+ may both give white precipitates with anion X^-, but only A^+ will give a white precipitate with Y^-. Thus, if both X^- and Y^- give white precipitates with an unknown, A^+ is present. If a precipitate occurs only with X^- but not Y^-, then B^+ is present. This behavior is summarized in Table 5-1.

Table 5-1. Behavior of hypothetical cations A^+ and B^+ with test reagents X^- and Y^-

Ion	Test reagent	
	X^-	Y^-
A^+	White precipitate	White precipitate
B^+	White precipitate	Clear solution

Analysis of Cations

For this experiment, the anion test reagents will be mixed with each of the known cation solutions. After studying the results of these tests, you will run additional tests to identify an unknown cation. The unknown will contain one of these seven cations.

Mn^{2+}	manganese(II) ion	manganese ion
Cu^{2+}	copper(II) ion	cupric ion
Fe^{3+}	iron(III) ion	ferric ion
Co^{2+}	cobalt(II) ion	cobaltous ion
Zn^{2+}	zinc(II) ion	zinc ion
Mg^{2+}	magnesium(II) ion	magnesium ion
Ba^{2+}	barium(II) ion	barium ion

The following reagents will be used to test for the presence of the cations.

$(NH_4)_2S$	ammonium sulfide
$(NH_4)_2SO_4$	ammonium sulfate
NH_4SCN	ammonium thiocyanate
$NaOH$	sodium hydroxide

Most chemical reactions may be assigned to one of these categories: precipitate formation, complex ion formation, gas evolution, and oxidation-reduction. Gas evolution and oxidation-reduction will be illustrated in later experiments.

Precipitate formation is easy to verify visually. When two clear aqueous solutions are mixed, a cloud of small solid particles forms and usually sinks to the bottom of the reaction vessel. The color of the precipitate should always be recorded.

Many cations form precipitates with sodium hydroxide (NaOH). As an example

$$Fe(NO_3)_3 + 3\ NaOH \longrightarrow Fe(OH)_3 + 3\ NaNO_3$$

This equation could also be written

$$Fe^{3+}(aq) + 3\ NO_3^-(aq) + 3\ Na^+(aq) + 3\ OH^-(aq) \longrightarrow Fe(OH)_3(s) + 3\ Na^+(aq) + 3\ NO_3^-(aq)$$

Note that the sodium ions (Na^+) and the nitrate ions (NO_3^-) do not participate in the reaction. They are "spectator ions." By removal of the spectator ions, a net ionic equation may be written.

$$Fe^{3+}(aq) + 3\ OH^-(aq) \longrightarrow Fe(OH)_3(s)$$

Cobalt forms complex ions such as $Co(NH_3)_6^{2+}$, $Co(SCN)_4^{2-}$, $Co(H_2O)_6^{2+}$, and $CoCl_4^{2-}$. The formation of a complex ion is often signaled by a change of color. In solutions containing low concentrations of chloride ion, the pink hexaaquo cobalt ion, $Co(H_2O)_6^{2+}$, predominates. At a larger chloride concentration, the blue tetrachloro cobalt complex, $CoCl_4^{2-}$, is formed.

PROCEDURE

Perform the following tests on the seven known cation solutions, study the results, and then perform each test on the unknown. **Always mix reagents thoroughly.** Look for the following signs of a chemical reaction: gas evolution, precipitate formation, a change in color.

Part A—Reactions of the Metal Ions with Sulfate Ions

Into each of seven clean labeled test tubes (16 × 150 mm) place about 10 mL of distilled water. Add 10 drops of 0.1 M manganese(II) nitrate, $Mn(NO_3)_2$, into tube 1, 10 drops of 0.1 M copper(II) nitrate, $Cu(NO_3)_2$, into tube 2, etc. Note the color of each solution. To each tube add 10 drops of 2 M ammonium sulfate, $(NH_4)_2SO_4$, solution. Mix thoroughly and carefully observe each tube. Record your observations on the Report Sheet. Write the formulas for all of the insoluble sulfates.

Part B—Reactions of Metal Ions with Sulfide Ions

Into each of seven clean labeled test tubes (16 × 150 mm) add 10 mL of distilled water and 10 drops of each metal ion solution as before. Now add 10 drops of 1 M ammonium sulfide $[(NH_4)_2S]$ solution. Mix thoroughly and carefully observe each tube. Record your observations on the Report Sheet. Write the formulas for all of the insoluble sulfides.

Part C—Reactions of Metal Ions with Thiocyanate Ions

Into each of seven clean labeled test tubes containing 10 mL water and 10 drops of each metal cation, add 10 drops of 2 M ammonium thiocyanate (NH_4SCN) solution. Mix thoroughly and carefully observe each tube for evidence of a reaction. Record your observations on the Report Sheet.

Part D—Reactions of Metal Ions with Hydroxide Ions

Into the seven clean labeled test tubes containing 10 mL of water and 10 drops of each metal cation, add 3 drops of 4 M sodium hydroxide (NaOH) solution. Mix thoroughly and carefully observe each tube for evidence of a reaction. Record your observations on the Report Sheet.

Part E—Unknown

After completing the tests on the known solutions, ask your instructor to approve your results and to issue an unknown to you. Each unknown contains a single metal ion. Perform each test using the procedures given above. Remember for each test to replace the known cation solution with your assigned unknown. Compare the results of each of the four tests on the seven known cation solutions with those for your unknown. Determine the identity of your unknown cation.

Report Sheet

_____   _____   _____
Name   Instructor/Section   Date

In the top space write the formula for each compound formed when the reagent anions are mixed with the ions in the cation column. Use the lower space to describe any visible change, i.e., yellow precipitate formed, solution turned red, etc. If no visible change occurs, write "N.R." in both spaces.

		Reagent added			
Tube	Cation	SO_4^{2-}	S^{2-}	SCN^-	OH^-
1	Mn^{2+}	_____	_____	_____	_____
		_____	_____	_____	_____
2	Cu^{2+}	_____	_____	_____	_____
		_____	_____	_____	_____
3	Co^{2+}	_____	_____	_____	_____
		_____	_____	_____	_____
4	Zn^{2+}	_____	_____	_____	_____
		_____	_____	_____	_____
5	Mg^{2+}	_____	_____	_____	_____
		_____	_____	_____	_____
6	Ba^{2+}	_____	_____	_____	_____
		_____	_____	_____	_____
7	Fe^{3+}	_____	_____	_____	_____
		_____	_____	_____	_____
Unknown		_____	_____	_____	_____
		_____	_____	_____	_____

Answer the questions on the next page.

QUESTIONS

1. Ammonia gas is formed when sodium hydroxide is added to ammonium ions in aqueous solution. The odor of household ammonia is that of ammonia gas. Look at the list of reagents for this experiment and fill in a chart for the results expected when each of the test anions, SO_4^{2-}, S^{2-}, SCN^- and OH^- is added to ammonium ion.

2. Knowing that calcium, Ca, is in the same family of the periodic chart with Mg and Ba, predict whether or not $Ca(OH)_2$ and $Ca(SCN)_2$ are soluble in water.

3. Silver chloride is insoluble, but cobalt chloride and barium chloride are soluble. Barium sulfate and silver sulfate are insoluble but cobalt sulfate is soluble. Prepare a chart of solubilities and indicate how sulfate ion and chloride ion could be used as test reagents to determine the presence of Co^{2+}, Ba^{2+}, or Ag^+ in an unknown.

EXPERIMENT

6   Analysis of Anions

APPARATUS

Test tube rack, 16 × 150 mm test tubes, medicine droppers, gummed labels.

REAGENTS

0.1 M solutions in dropping bottles of sodium chloride (NaCl), sodium iodide (NaI), sodium nitrate (NaNO$_3$), sodium sulfate (Na$_2$SO$_4$), sodium carbonate (Na$_2$CO$_3$), silver nitrate (AgNO$_3$), and barium nitrate [Ba(NO$_3$)$_2$]. Solutions in dropping bottles of 6 M nitric acid (HNO$_3$), 1 M ammonium sulfide [(NH$_4$)$_2$S], and 4 M sodium hydroxide (NaOH). Unknown solutions.

INTRODUCTION

Solutions of dissolved ionic materials may be analyzed to determine the presence or absence of specific anions. An analysis done to establish only the identity of the ions is qualitative analysis. An analysis to determine the amounts of each ion is quantitative analysis. In this experiment you will do a qualitative analysis of an anion in an unknown.

Qualitative analysis is based on using chemical tests that will give different results for each of several possible unknown materials. Ideally if a single reagent would react with only a specific anion, the observation of the expected reaction would unambiguously identify the unknown. The absence of a reaction would mean the absence of that specific anion. However, few reagents give specific tests for one anion of all the possible ions.

Qualitative analysis is then based on determining the response of an ion to several reagents. No two ions behave in exactly the same way when exposed to a variety of reagents. For example anions X^- and Y^- may both give white precipitates with cation A^+, but only X^- will give a white precipitate with B^+. Thus, if both A^+ and B^+ give white precipitates with an unknown, X^- is present. If a precipitate occurs only with A^+ but not B^+, then Y^- is present. This behavior is summarized in Table 6-1.

Table 6-1. Behavior of hypothetical anions X^- and Y^- with test reagents A^+ and B^+

Ion	Test reagent	
	A^+	B^+
X^-	White precipitate	White precipitate
Y^-	White precipitate	Clear solution

Analysis of Anions

For this experiment, the test reagents will be mixed with each of the known anion solutions. After studying the results of these tests, additional tests are run to identify an unknown anion. The unknown will contain one of these seven anions.

Cl^-	chloride ion	CO_3^{2-}	carbonate ion
I^-	iodide ion	S^{2-}	sulfide ion
NO_3^-	nitrate ion	OH^-	hydroxide ion
SO_4^{2-}	sulfate ion		

The following reagents will be used to test for the presence of the anions.

$AgNO_3$	silver nitrate
$Ba(NO_3)_2$	barium nitrate
HNO_3	nitric acid

Most chemical reactions may be assigned to one of these categories: precipitate formation, complex ion formation, gas evolution, and oxidation-reduction. Gas evolution can be produced from many anions. The nature of the gas can be used to identify the anion.

Upon adding acid, many anions evolve gases. In this experiment nitric acid will be added to test for gas evolution. Anions that produce gases upon acidification include those listed in Table 6-2. Note also the characteristics of the gases evolved.

Table 6-2

Anion	Gas evolved upon acidification	Properties of gas
CO_3^{2-}	CO_2 (carbon dioxide)	Colorless Odorless Noncombustible
SO_3^{2-}	SO_2 (sulfur dioxide)	Colorless Sharp odor Noncombustible
S^{2-}	H_2S (hydrogen sulfide)	Colorless Rotten egg odor Combustible

As in the experiment on cation analysis, precipitate formation is also an excellent method of identifying anions. Two cation test reagents will be employed to test for precipitation of anions: silver nitrate ($AgNO_3$) and barium nitrate [$Ba(NO_3)_2$]. Nitric acid (HNO_3) will also be added to test for gas evolution.

Hydroxide is a special anion because water is formed when acid is added. An insoluble hydroxide salt can dissolve in acid to form water. Other insoluble salts may also be soluble in acids if the anion reacts with the acid. For example, barium sulfite ($BaSO_3$) is insoluble but dissolves when nitric acid is added.

$$BaSO_3(s) + 2\,HNO_3(aq) \longrightarrow Ba(NO_3)_2(aq) + H_2O + SO_2(g)$$

PROCEDURE

Before performing this experiment study Section C, "Testing for Odors (Wafting)," on pages 8-9 of this manual.

Perform the following tests on the seven known anion solutions, study the results, and then perform each test on the unknown. Always mix reagents thoroughly. Always look for the following signs of a chemical reaction: gas evolution, precipitate formation, precipitate dissolution, a change in color.

CAUTION: Nitric acid (HNO₃) should not be spilled or splashed out of the test tube. Promptly wash any acid spilled on your skin with large quantities of water. Immediately inform your instructor about your accident.

CAUTION: Avoid spilling silver nitrate (AgNO₃) on your skin. Although the skin discoloration (black) is not a health hazard, the results may be considered unsightly. When new skin is formed, the affected skin will be discarded.

Part A—Reactions of Anions with Silver Ions

Into each of seven clean labeled test tubes (16 × 150 mm) place about 10 mL of distilled water. Then add 10 drops of 0.1 M sodium chloride (NaCl) into tube 1, 10 drops of 0.1 M sodium iodide (NaI) into tube 2, etc. Add 1 mL (20 drops) of 0.1 M silver nitrate (AgNO₃), to each test tube. Shake each tube to insure proper mixing. Record your observations in column 1 on the Report Sheet and save each tube for Part B.

Part B—Reactions of Anions with Silver and Hydrogen Ions

Add 1 mL (20 drops) of 6 M nitric acid (HNO₃) to each mixture prepared in Part A. Shake each tube to insure proper mixing. Note carefully any gas bubbles. Gently waft gas vapors toward your nose and note any odor. Record these observations in column 2 on the Report Sheet.

Part C—Reactions of Anions with Barium Ions

Repeat Part A, replacing the silver nitrate with 0.1 M barium nitrate, Ba(NO₃)₂. Shake each tube to insure complete mixing and record your observations in column 3 of the Report Sheet. Save the tubes for Part D.

Part D—Reaction of Anions with Barium and Hydrogen Ions

Add 1 mL (20 drops) of 6 M nitric acid (HNO₃) to each of the seven mixtures prepared in Part C. Agitate each tube and note carefully any gas bubbles. Gently waft gas vapors toward your nose and note any odor. Record these observations in column 4 on the Report Sheet.

Part E—Unknown

After completion of the tests on the known solutions, ask your instructor to approve your results and then to issue to you an unknown. Each unknown contains a single anion. Perform each test using the procedures in Parts A through D. Remember for each test to replace the known anion solution with your assigned unknown. Compare the results of each of the four tests on the seven known anion solutions with those for your unknown. What is your unknown anion?

Report Sheet

Name	**Instructor/Section**	**Date**	

In the top space write the formula for each compound formed when the reagent ions are mixed with the ions in the anion column. Use the bottom space to describe any visible change, i.e., yellow precipitate formed, solution turned red, etc. If no visible change occurs, write "N.R." in both spaces.

Reagent added

Tube	Anion	Ag^+ alone	Ag^+ and H^+	Ba^{2+} alone	Ba^{2+} and H^+
1	Cl^-	_____	_____	_____	_____
		_____	_____	_____	_____
2	I^-	_____	_____	_____	_____
		_____	_____	_____	_____
3	NO_3^-	_____	_____	_____	_____
		_____	_____	_____	_____
4	SO_4^{2-}	_____	_____	_____	_____
		_____	_____	_____	_____
5	CO_3^{2-}	_____	_____	_____	_____
		_____	_____	_____	_____
6	S^{2-}	_____	_____	_____	_____
		_____	_____	_____	_____
7	OH^-	_____	_____	_____	_____
		_____	_____	_____	_____
Unknown		_____	_____	_____	_____
		_____	_____	_____	_____

Answer the questions on the next page.

QUESTIONS

1. Using the data in Table 6-1 for the behavior of SO_3^{2-} and your experience with SO_4^{2-} in this experiment, devise a test for differentiating the two. Assume you have an unknown containing either SO_3^{2-} or SO_4^{2-}. What substance would you add to find out which is present? Explain your answer.

2. Knowing that chlorine, bromine, and iodine are in the same family of the periodic chart, predict whether or not silver bromide is soluble in water.

3. Magnesium phosphate is insoluble in water but the solid dissolves in HNO_3. Write an equation for the reaction.

4. Cadmium carbonate is insoluble in water. Write an equation for the expected reaction of cadmium carbonate with H_2SO_4.

EXPERIMENT
7
Chemical Reactions

APPARATUS

10 mL graduated cylinder, medicine dropper, micro spatula, 16 × 150 mm test tubes, test tube rack.

REAGENTS

Solids: Sodium hydrogen carbonate ($NaHCO_3$), sodium hydrogen sulfite ($NaHSO_3$), iron(II) sulfide (FeS), magnesium ribbon (Mg) cut in 2-3 cm strips, litmus paper, matches.

Solutions: 6 *M* hydrochloric acid (HCl), 6 *M* aqueous ammonia (NH_3), 0.1 *M* barium chloride ($BaCl_2$), 0.1 *M* sodium chromate (Na_2CrO_4), 0.1 *M* sodium sulfate (Na_2SO_4), 0.1 *M* iron(III) chloride ($FeCl_3$), 0.1 *M* ammonium thiocyanate (NH_4SCN), 0.1 *M* copper(II) sulfate ($CuSO_4$).

INTRODUCTION

A chemical reaction involves the conversion of one or more substances called reactants into one or more substances called products. When a chemical reaction occurs, there are frequently, but not always, changes that can be detected by your physical senses. In this experiment you will use your senses of sight, smell, and touch to determine when a chemical reaction occurs. You are not to use your sense of taste in this experiment.

WARNING: Never taste a chemical in the laboratory.

One of the ways you can sense that a chemical reaction has occurred is by observing the formation of a precipitate. A precipitate results when the product is insoluble in the solvent. The insoluble product may be small particles that appear as a cloudy suspension or large particles that rapidly settle to the bottom of the test tube. In describing a precipitate, you should also note its color.

A second way of determining that a reaction has occurred is by observing the formation of a gas. Gas evolution is evidenced by the formation of bubbles in the reaction mixture when the gaseous product is insoluble in the solvent. Gas evolution may occur from the decomposition of unstable acids. For example, carbonic acid (H_2CO_3) will form carbon dioxide and water on heating. Similarly, sulfurous acid (H_2SO_3) forms sulfur dioxide (SO_2) and water.

$$H_2CO_3(aq) \longrightarrow H_2O + CO_2(g)$$

$$H_2SO_3(aq) \longrightarrow H_2O + SO_2(g)$$

Most gases are colorless and odorless. However, some gases have distinctive odors. Thus, the sense of smell can be used to identify these gases.

All experiments that may produce gases must be performed in a fume hood.

In order to test for the odor of the gas, fan the vapors gently toward your nose. Do this by waving your hand gently over the open end of the test tube. See Section C, "Testing for Odors (Wafting)," and Figure 5 (pages 8-9).

CAUTION: Do not place the test tube directly under your nose.

A third way of determining that a reaction has occurred is by observing a change in the color of the solution. Color changes can be quite vivid in oxidation-reduction reactions. One substance loses electrons and is oxidized while a second substance gains electrons and is reduced. If a substance such as a metal ion can exist in two or more oxidation states and these oxidation states have different colors, then a chemical change can be easily observed. The most commonly encountered colored substances are transition metal ions which form complex ions, which have a variety of colors. Thus, a conversion from one complex ion into another can be accompanied by a change in color.

A fourth way to observe a chemical reaction is to detect a change in the temperature of the reaction solution. When energy is released by chemicals as a result of a chemical reaction, there is an increase in the temperature of the reaction mixture. Such a reaction is called exothermic. If energy is absorbed from the surroundings when a reaction occurs, the temperature of the reaction mixture will decrease. Such a reaction is called endothermic. The sense of touch then can be used to determine whether a reaction has occurred if sufficient energy is required or is released. This is done by touching the test tube containing the reaction mixture to determine if it feels warmer or colder..

PROCEDURE

Before performing any part of this experiment, study the following sections in this manual: A, "Handling Liquid Reagents," B, "Handling Solid Reagents," and C, "Testing for Odors (Wafting)," on pages 6-9.

Wash ten 16 × 150 mm test tubes and rinse each one thoroughly with distilled water. These tubes will be used for Parts A through J.

The quantities of solid material used in this experiment need only be approximate. Use the micro spatula to add the solids to the test tube. Rinse and wipe the spatula after each use.

Drops of solution are delivered with a medicine dropper. The dropper must be cleaned and rinsed with distilled water after each use. To do this properly, the rubber bulb must be removed from the dropper. For 6 M ammonia and 6 M hydrochloric acid, use the dispensing containers that deliver dropwise quantities.

The 5 mL volumes of solutions required in some parts of this experiment may be added from a 10 mL graduated cylinder. Wash this cylinder after each use and rinse it several times with distilled water. Agitation of a test tube to mix reagents should be done gently so that the contents are not splashed out. Hold the test tube near its top with a finger and thumb. Tap the tube on the side near the bottom with a finger of the other hand.

WARNING: Always aim the mouth of a test tube away from yourself and from your neighbors so they are not splashed with chemicals that may spurt from the tube.

In addition to odor, other properties that will be used in this experiment to identify a gas are listed in Table 7-1.

Table 7-1

Gas	Color	Odor	Match	Litmus
Hydrogen, H_2	None	None	Explodes	No effect
Oxygen, O_2	None	None	Burns brightly	No effect
Hydrogen sulfide, H_2S	None	Rotten eggs	Burns	Red
Sulfur dioxide, SO_2	None	Sharp	—	Red
Carbon dioxide, CO_2	None	None	—	Red
Ammonia, NH_3	None	Ammonia	—	Blue

A reaction in which a gas is evolved may have to be run a second or third time in order to complete all of the observations, sight, smell, touch, match, and litmus. Parts A through D must be done in a fume hood because unpleasant-smelling gases may be produced.

CAUTION: 6 *M* hydrochloric acid can cause severe skin burns. If you do spill acid on your skin, wash the affected areas at once with large quantities of water. Notify your instructor and seek a medical evaluation of the chemical burn.

The litmus test is performed as follows. A piece of red litmus paper and a piece of blue litmus paper are moistened with distilled water and each is held separately inside the top of the test tube while gas evolution is occurring. Do not allow the paper to touch the test tube since it will react with the liquid reaction solution. One of the following results will be obtained.

1. Neither paper changes color.
2. Both papers are blue.
3. Both papers are red.

Part A (Hood)

Carefully add 5 mL of 6 *M* hydrochloric acid (HCl) to a graduated cylinder and transfer the acid to a 16 × 150 mm test tube. Add a 2-3 cm strip of magnesium ribbon to the hydrochloric acid in the test tube. Record your observations (sight, smell, and touch) on the Report Sheet.

For safety, secure the test tube in a test tube rack.

Add another strip of magnesium and bring a burning match to the open end of the test tube. Record your observations. Test the gas with wet litmus paper, and record your observations. Write the products of the reaction between magnesium (Mg) and hydrochloric acid (HCl) on the Report Sheet.

Part B (Hood)

Place a spatulaful of solid sodium hydrogen carbonate ($NaHCO_3$) in a test tube. Slowly add 6 drops of 6 *M* hydrochloric acid (HCl) to the test tube. Record your observations. Repeat the experiment, but this time bring a lighted match to the mouth of the test tube. Record your observations. Test the gas with wet litmus paper, and record your observations. Write the products of the reaction between sodium hydrogen carbonate and hydrochloric acid on the Report Sheet.

Part C (Hood)

Place a spatulaful of solid sodium hydrogen sulfite ($NaHSO_3$) in a test tube. Slowly add 6 drops of 6 M hydrochloric acid to the test tube. Record your observations. Test the gas with litmus paper, and record your observations. Do not use a burning match in this part!

Part D (Hood)

Place a small piece (about the size of a grain of rice) of iron(II) sulfide (FeS) in a test tube. Add 6 drops of 6 M hydrochloric acid. Record your observations. Test the gas with wet litmus paper, and record your observations. Do not use a burning match in this part!

Part E

Place 5 mL of copper(II) sulfate ($CuSO_4$) solution in a test tube. Add an iron tack to the test tube. (The iron tack should be cleaned by momentary immersion in 6 M HCl and then rinsed with distilled water prior to use.) Set the test tube aside and record your observations after 10-15 minutes.

Part F

Place 5 mL of barium chloride ($BaCl_2$) solution in a test tube. Add 6 drops of sodium chromate (Na_2CrO_4) solution to the test tube. Mix the reagents by swishing the test tube. Record your observations.

Part G

Place 5 mL of barium chloride ($BaCl_2$) solution in a test tube. Add 5 mL of sodium sulfate (Na_2SO_4) solution to the test tube. Agitate the test tube. Record your observations.

Part H

Place 3 mL of iron(III) chloride ($FeCl_3$) solution in a test tube. Add 1 or 2 drops of ammonium thiocyanate (NH_4SCN) solution to the test tube. Agitate the test tube. Record your observations. The formula of the colored species is $FeSCN^{2+}$.

Part I

Place 5 mL of the copper(II) sulfate solution in a test tube. Add 20 drops of 6 M ammonium hydroxide (NH_4OH) solution (ammonium hydroxide is another name for a solution of ammonia, NH_3, in water). Agitate the test tube. Record your observations. The formula of the colored species is $Cu(NH_3)_4^{2+}$.

Part J (Hood demonstration by laboratory instructor)

Use crucible tongs to grasp a piece of magnesium ribbon about 2-3 cm in length. Hold it in a burner flame until it ignites. Record your observations.

CAUTION: Do not stare directly at the burning magnesium ribbon.

Report Sheet

	Name		Instructor/Section		Date

Part	Sight	Smell	Touch	Match[1]	Litmus[2]	Product[3]
A	_____	_____	_____	_____	_____	_____
	_____	_____	_____	_____	_____	
B	_____	_____	_____	_____	_____	_____
	_____	_____	_____	_____	_____	
C	_____	_____	_____		_____	_____
	_____	_____	_____		_____	
D	_____	_____	_____		_____	_____
	_____	_____	_____		_____	
E	_____	_____	_____			_____
	_____	_____	_____			
F	_____	_____	_____			_____
	_____	_____	_____			
G	_____	_____	_____			_____
	_____	_____	_____			
H	_____	_____	_____			_____
	_____	_____	_____			
I	_____	_____	_____			_____
	_____	_____	_____			
J	_____	_____	_____			_____
	_____	_____	_____			

1   The match test is to be performed for Parts A and B only.
2   The litmus test is to be performed for Parts A through D only.
3   The column of chemical formulas for each major product must be filled out before you leave the laboratory.

Chemical Reactions

Complete and balance these chemical equations, which describe what occurred in each part of the experiment.

A $\qquad$ $HCl + Mg \longrightarrow MgCl_2 +$ _____

B $\qquad$ $HCl + NaHCO_3 \longrightarrow NaCl + H_2O +$ _____

C $\qquad$ $HCl + NaHSO_3 \longrightarrow NaCl + H_2O +$ _____

D $\qquad$ $HCl + FeS \longrightarrow FeCl_2 +$ _____

E $\qquad$ $Fe + CuSO_4 \longrightarrow FeSO_4 +$ _____

F $\qquad$ $BaCl_2 + Na_2CrO_4 \longrightarrow NaCl +$ _____

G $\qquad$ $BaCl_2 + Na_2SO_4 \longrightarrow NaCl +$ _____

H $\qquad$ $FeCl_3 + NH_4SCN \longrightarrow NH_4Cl + Cl^- +$ _____

I $\qquad$ $CuSO_4 + 4\,NH_3 \longrightarrow$ _____ $+ SO_4^{2-}$

J $\qquad$ $Mg + O_2 \longrightarrow$ _____

Attach the questions sheet.

82

QUESTIONS

1. How is a chemical change distinguished from a physical change?

2. List four kinds of observations that can lead you to conclude that a chemical reaction has occurred.

3. How small a change in temperature can you detect by touch? How could smaller changes in temperature be detected?

4. What would you expect to observe if $KHCO_3$ reacted with H_2SO_4? What test would you perform to identify the product?

5. What difference, if any, would be observed if $CuCl_2$ were used in Part E rather than $CuSO_4$?

6. What chemical species is responsible for the color of the precipitate in Part F? What can be concluded about the color of the barium ion (Ba^{2+}) based on observations in Parts F and G?

7. Complete the following table by combining the cation and anion for each empty square. Each compound must be electrically neutral. Name each compound.

	Cl^-	S^{2-}	SO_4^{2-}	CrO_4^{2-}	OH^-
Na^+					
K^+			K_2SO_4 potassium sulfate		
Ba^{2+}				$BaCrO_4$ barium chromate	
Fe^{3+}					
Cu^{2+}					$Cu(OH)_2$ copper(II) hydroxide
NH_4^+		$(NH_4)_2S$ ammonium sulfide			

EXPERIMENT

8   Empirical Formula of an Oxide of Tin

APPARATUS

Crucible (size 0) with cover, clay triangle, crucible tongs, ceramic mat, ring stand, iron ring, clamp, medicine dropper, Meker burner, matches, rubber tubing, beaker.

REAGENTS

1 g of 30 mesh granulated tin, 5 mL of 10 M nitric acid (HNO_3).

INTRODUCTION

The composition of a compound is independent of its source, providing it has been purified and separated from all other compounds. The composition can be determined by chemically breaking the compound into its component elements. The same elements can be recombined in the proper amounts, dictated by the composition of the compound, to produce the compound. These facts are summarized in the law of definite proportions: When elements combine to form a compound, they do so in definite proportions by mass.

An illustration of the law of definite proportions is the combining of iron and sulfur to form iron sulfide. If 55.85 g of iron is heated with 32.06 g of sulfur, then 87.91 g of iron sulfide results. No iron or sulfur remains. If 55.85 g of iron is heated with 40.00 g of sulfur, only 87.91 g of iron sulfide results. In this case 7.94 g of sulfur remains because this quantity is in excess of the amount required to form iron sulfide. The excess sulfur can be burned off to produce gaseous sulfur dioxide, and then the 87.91 g of pure iron sulfide is obtained.

In this experiment, a fixed amount of tin is reacted with an excess of nitric acid to form an oxide of tin. The excess nitric acid will then be decomposed to gaseous products leaving only pure tin oxide. From the known masses of tin and of the tin oxide, you can calculate the mass of oxygen that reacted. The ratio of the numbers of moles of tin and oxygen atoms can be calculated. The empirical formula will be established from this ratio.

For this experiment you will be using 10 M nitric acid, a concentrated solution of a strong acid. Exercise caution whenever working with this acid. As a precautionary measure, make it a habit to wash your hands after handling the acid bottle and other containers. First aid for acid spills consists of immediate washing with large quantities of cold running water.

The reaction of tin and nitric acid is vigorous and gives off large quantities of steam and nitrogen dioxide (NO_2).

CAUTION: Nitrogen dioxide is a poisonous red-brown gas. For this reason the reaction is to be carried out only in the hood.

PROCEDURE

Study Section N, ''Use of the Balances,'' on pages 27-29 of this manual. To avoid errors you should use the same analytical balance for all weighings in this experiment. Study also Sections E, "The Gas Burner" (pages 11-12), and G, "Use of the Crucible" (pages 14-15). You will have to know how to adjust the Meker burner, use a crucible, fire a crucible, and ignite a crucible with its contents.

Obtain a porcelain crucible with a cover and wash both with soap and water. After drying with a towel, place the crucible with the cover ajar on a clay triangle supported by an iron ring (see Figure 15 on page 15 of this manual). Adjust the gas burner to a cool flame and heat the crucible gently for a few minutes. Then adjust the burner to give a very hot flame. Open the gas jet all the way and then adjust the mixture control at the bottom of the burner until a blue cone appears and a hissing sound is heard. Place the crucible in the upper part of the blue cone. A properly adjusted burner will make the crucible glow orange-red after only a few minutes.

After heating the glowing crucible for 10 minutes, turn off the burner. From this point on, handle the crucible and cover only with tongs, to prevent the moisture of your hands from being deposited on them. Place the crucible on a ceramic mat to cool for about 10 minutes (see Figure 17 on page 15 of this manual). During cooling, the lid should be placed tightly on the crucible. After the crucible and lid have cooled to room temperature, determine their mass with the analytical balance. Record this mass on the Report Sheet.

Reheat the crucible with the cover ajar for 5 minutes. After allowing time for the apparatus to cool to room temperature, weigh the crucible and cover on the analytical balance. If the second weight differs from the first weight by more than 0.005 g, repeat the heating process until subsequent weighings agree within 0.005 g.

CAUTION: Nitric acid is a strong corrosive acid that will cause severe skin burns. Immediately wash any affected areas with large quantities of water. Notify your instructor and seek a medical evaluation of the chemical burn.

Using the top-loading balance, weigh from 1.0 to 1.1 g of 30 mesh granulated tin and add it to the crucible. Replace the cover and reweigh on the analytical balance. Record this mass on the Report Sheet. Place the crucible in the hood, and add 10 M nitric acid dropwise at the rate of 1 drop/second until 50 drops have been added. The reaction is vigorous and may throw some of the oxide onto the inside walls of the crucible. In this case, use your last 5-10 drops of the nitric acid to wash down the sides of the crucible.

When the reaction has subsided, use tongs to place the crucible on a clay triangle in the hood. Heat gently, then strongly, with a Meker burner until all of the brown fumes have been driven off. Place the cover, slightly ajar, on the crucible and heat strongly for 15 minutes. Allow the crucible, cover, and contents to cool to room temperature. Weigh the crucible, cover, and contents on the analytical balance. Record the mass. Reheat the crucible with the cover ajar for

5 minutes. After cooling to room temperature, reweigh the crucible, cover, and contents on the analytical balance. If the mass loss since the previous weighing exceeds 0.005 g, the contents of the crucible were not dry when originally weighed and the heating process should be repeated until the weight loss on successive heatings is less than 0.005 g.

CALCULATIONS

A student obtained the following data for the determination of the formula of tin oxide.

Mass of crucible and lid   = 22.923 g

Mass of crucible, lid, and tin   = 24.163 g

Mass of crucible, lid, and tin oxide   = 24.412 g

The formula is determined as follows:

1. Calculate the mass of tin metal used.

$$\text{Mass of tin} = (\text{mass of tin} + \text{crucible} + \text{lid}) - (\text{mass of crucible} + \text{lid})$$
$$= 24.163 \text{ g} - 22.923 \text{ g}$$
$$= 1.240 \text{ g}$$

2. Calculate the moles of tin metal used.

$$\text{Moles of tin} = \frac{\text{mass of tin}}{\text{atomic weight of tin}}$$
$$= \frac{1.240 \text{ g}}{118.69 \text{ g/mole}}$$
$$= 1.045 \times 10^{-2} \text{ mole}$$

3. Calculate the mass of oxygen that combined with the tin.

$$\text{Mass of oxygen} = (\text{mass of crucible} + \text{cover} + \text{tin oxide}) - (\text{mass of crucible} + \text{cover} + \text{tin})$$
$$= 24.412 \text{ g} - 24.163 \text{ g}$$
$$= 0.249 \text{ g}$$

4. Calculate the moles of oxygen that combined.

$$\text{Moles of oxygen} = \frac{\text{mass of oxygen}}{\text{atomic weight of oxygen}}$$
$$= \frac{0.249 \text{ g}}{15.9994 \text{ g/mole}}$$
$$= 1.556 \times 10^{-2} \text{ mole}$$

5. Calculate the ratio of the number of moles of oxygen to the number of moles of tin.

$$\frac{\text{Moles of oxygen}}{\text{Moles of tin}} = \frac{1.556 \times 10^{-2} \text{ mole}}{1.045 \times 10^{-2} \text{ mole}}$$

$$= 1.489$$

This result gives the formula for the oxide of tin as

$SnO_{1.489}$

Rounding to the closest whole number ratio, we obtain

Sn_2O_3

Report Sheet

_____ _____ _____
Name   Instructor/Section   Date

Mass of crucible and cover (g)   (1a) _____

                                                  (1b) _____

                                                  (1c) _____

Mass of crucible, cover, and tin (g)   (2) _____

Mass of tin (g)   (3) _____

Mass of crucible, cover, and oxide (g)   (4a) _____

                                                  (4b) _____

                                                  (4c) _____

Mass of oxygen (g)   (5) _____

Moles of tin atoms in oxide   (6) _____

Moles of oxygen atoms in oxide   (7) _____

Ratio of moles of oxygen to moles of tin   (8) _____

Empirical formula of tin oxide   (9) _____

Show your calculations of the above quantities.
Answer the questions on the next page.

Empirical Formula of an Oxide of Tin

QUESTIONS

1. In this experiment, how would your determination of the empirical formula be affected if the oxide were not heated to dryness?

2. How would your determination of the empirical formula be affected if insufficient nitric acid were added and some tin remained mixed with the oxide?

3. Suppose some of the oxide splattered out of the crucible. How would this affect your calculated ratio of the number of moles of oxygen to the number of moles of tin?

4. Suppose that the tin used in this experiment had a thin oxide coating. How would this affect your calculated ratio of the number of moles of oxygen to the number of moles of tin?

5. Suppose some material adhered to the bottom of your crucible containing the tin oxide while it was cooling. How would this affect your calculated ratio of the number of moles of oxygen to the number of moles of tin?

EXPERIMENT
9

Synthesis of Sodium Carbonate

APPARATUS

Crucible (size 0) with cover, clay triangle, crucible tongs, ceramic mat, ring stand, iron ring, clamp, Meker burner, matches, rubber tubing.

REAGENTS

Solid sodium hydrogen carbonate ($NaHCO_3$).

INTRODUCTION

The purpose of this experiment is to make sodium carbonate by heating sodium hydrogen carbonate in a crucible. When heated over 100°C, sodium hydrogen carbonate decomposes according to the following equation.

$$2 \ NaHCO_3 \xrightarrow{\Delta} Na_2CO_3 + H_2O(g) + CO_2(g)$$

You will calculate the percent yield for your preparation of sodium carbonate. The only reactant is sodium hydrogen carbonate; thus the only factors that limit the yield of product obtained are the care that you take not lose any product. The mass of the product calculated from the mass of the reactant used is the theoretical yield. The actual yield is the mass of product actually obtained in the experiment. The percent yield is calculated as

$$\frac{\text{Actual yield}}{\text{Theoretical yield}} \times 100\% = \text{percent yield}$$

PROCEDURE

Before performing this experiment study the following sections of this manual: E, "The Gas Burner" (pages 11-12); G, "Use of the Crucible" (pages 14-15); and N, "Use of the Balances" (pages 27-29).

Obtain a porcelain crucible with a cover and wash both with soap and water. After drying with a towel, place the crucible with the cover ajar on a clay triangle supported by an iron ring (see

Synthesis of Sodium Carbonate

Figure 15 on page 15 of this manual). Adjust the burner to a cool flame and heat the crucible gently for a few minutes. Then adjust the burner to give a very hot flame. Open the gas jet all the way and then adjust the mixture control at the bottom of the burner until a blue cone appears and a hissing sound is heard. Place the crucible in the upper part of the blue cone. A properly adjusted burner will make the crucible glow orange-red after only a few minutes.

After heating the glowing crucible for 10 minutes, turn off the burner. From this point on, handle the crucible and cover only with tongs, to prevent the moisture of your hands from being deposited on them. Place the crucible on a ceramic mat to cool for about 10 minutes (see Figure 17 on page 15 of this manual). During cooling, the lid should be placed tightly on the crucible.

Weigh the crucible on the analytical balance. Repeat this heating procedure followed by cooling until two subsequent weighings agree within 0.005 g. After the final weighing add approximately 3 g of sodium hydrogen carbonate to the crucible and record the combined weights to the nearest 0.001 g.

Heat the crucible containing the sodium hydrogen carbonate in the same way the empty crucible was heated. Repeat heating, cooling, and weighing of the crucible with contents until the weight is constant within 0.005 g.

If sufficient time is available, repeat the experiment with another sample of sodium hydrogen carbonate.

CALCULATIONS

Suppose that 3.638 g of sodium hydrogen carbonate ($NaHCO_3$) were heated, producing 2.000 g of sodium carbonate (Na_2CO_3). What is the percent yield?

1. Calculate the number of moles of $NaHCO_3$ in the sample.

$$\text{Number of moles} = n = \frac{\text{mass of sample}}{\text{molecular weight}} = \frac{3.638 \text{ g}}{84.00 \text{ g/mole}} = 4.331 \times 10^{-2} \text{ mole}$$

2. Calculate the theoretical yield (expected mass of Na_2CO_3). From the equation for the reaction we see that

$$\text{Moles of } Na_2CO_3 \text{ expected} = \frac{1}{2} \text{ (moles of } NaHCO_3 \text{ taken)}$$
$$= \frac{1}{2} \times (4.331 \times 10^{-2} \text{ mole})$$
$$= 2.165 \times 10^{-2} \text{ mole}$$

$$\text{Theoretical yield} = \text{mass of } Na_2CO_3 \text{ expected}$$
$$= 2.165 \times 10^{-2} \text{ mole} \times 106.00 \text{ g/mole}$$
$$= 2.294 \text{ g}$$

3. Determine by experiment the mass of Na_2CO_3 actually produced.

$$\text{Actual mass of } Na_2CO_3 = 2.000 \text{ g}$$

4. Calculate the percent yield.

$$\% \text{ yield} = \frac{\text{actual mass produced}}{\text{theoretical yield}} \times 100\% = \frac{2.000 \text{ g}}{2.294 \text{ g}} \times 100\% = 87.18\%$$

Report Sheet

	Name	Instructor/Section	Date

	I	II

Weight of empty crucible

    After first heating (g)   (1a) _____ _____

    After second heating (g)   (1b) _____ _____

    After third heating (g)   (1c) _____ _____

Weight of crucible and sodium hydrogen carbonate

    Before heating (g)   (2) _____ _____

Weight of sodium hydrogen carbonate (g)   (3) _____ _____

Moles of sodium hydrogen carbonate taken   (4) _____ _____

Weight of crucible and sodium carbonate

    After first heating (g)   (5a) _____ _____

    After second heating (g)   (5b) _____ _____

    After third heating (g)   (5c) _____ _____

Weight of sodium carbonate product (g)   (6) _____ _____

Expected yield of sodium carbonate (g)   (7) _____ _____

% yield   (8) _____ _____

Attach sheets with your calculations. Answer the questions on the next page.

Synthesis of Sodium Carbonate

QUESTIONS

1. How would you explain a percent yield of sodium carbonate greater than 100%?

2. Explain how you would calculate your result if initially the sodium hydrogen carbonate contained two molecules of water of hydration.

3. Explain how this experimental procedure could be used to determine the amount of sodium hydrogen carbonate in a sample of sodium hydrogen carbonate and sodium chloride, assuming that a 100% yield of sodium carbonate is obtained.

4. Calculate the theoretical yield of sodium hydrogen carbonate, $NaHCO_3$, if 20.3 g of sodium chloride, NaCl, are reacted with 15.7 g of ammonium carbonate, $(NH_4)_2CO_3$, according to the equation

$$NaCl + (NH_4)_2CO_3 \longrightarrow NaHCO_3(s) + NH_3(g) + NH_4Cl$$

EXPERIMENT

10   Water of Hydration

APPARATUS

Five 25 X 200 mm hard glass test tubes, crucible (size 0) with cover, clay triangle, ring stand, iron ring, clamp, crucible tongs, Meker burner, matches, rubber tubing.

REAGENTS

Sucrose, copper sulfate, sodium sulfate, potassium chloride, cobalt chloride, magnesium sulfate.

INTRODUCTION

Several observations are possible when a solid compound is heated. These include melting, irreversible decomposition, reversible decomposition, and nothing at all. For a solid hydrate (a compound containing water of hydration) the water of hydration can be driven off by heating; the result is the anhydrous compound, and the original hydrate can be regenerated simply by the addition of water. If the addition of water does not reproduce the original compound, the process is irreversible and the original compound was not a hydrate. Compounds containing water of hydration usually are written with the water separated by a dot; for example, $CuSO_4 \cdot 5H_2O$. This means for hydrated copper sulfate there are 5 moles of water bound to each mole of copper sulfate.

PROCEDURE

Before performing this experiment study the following sections of this manual: B, "Handling Solid Reagents" (pages 7-8); E, "The Gas Burner" (pages 11-12); F, "Heating Liquids" (pages 12-14); G, "Use of the Crucible" (pages 14-15); and N, "Use of the Balances" (pages 27-29).

Part A—Qualitative Study of Hydration

Place about 0.5 g of each solid compound listed as reagents into separate hard glass test tubes. Heat each in a burner flame and record your observations. If no water droplets condense onto the cooler upper wall of the test tube, the compound does not contain water and is not a hydrate. If water droplets condense onto the upper wall of the test tube, the water resulted either from irreversible decomposition or from the reversible loss of water of hydration. After allowing each residue to cool, add about 10 drops of distilled water. If the residue is not soluble in water, the compound decomposed irreversibly and is not a hydrate. Record also all color changes that occur.

Part B—Determination of Water Content in Hydrated Magnesium Sulfate

Obtain a porcelain crucible with a cover and wash both with soap and water. After drying with a towel, place the crucible with the cover ajar on a clay triangle supported by an iron ring (see Figure 15 on page 15 of this manual). Adjust the gas burner to a cool flame and heat the crucible gently for a few minutes. Then adjust the burner to give a very hot flame. Open the gas jet all the way and then adjust the mixture control at the bottom of the burner until a blue cone appears and a hissing sound is heard. Place the crucible in the upper part of the blue cone. A properly adjusted burner will make the crucible glow orange-red after only a few minutes.

After heating the glowing crucible for 10 minutes, turn off the burner. From this point on, handle the crucible and cover only with tongs, to prevent the moisture of your hands from being deposited on them. Place the crucible on a ceramic mat to cool for about 10 minutes (see Figure 17 on page 15 of this manual). During cooling, the lid should be placed tightly on the crucible. After the crucible and lid have cooled to room temperature, determine their mass with the analytical balance. Record this mass on the Report Sheet.

Repeat this procedure (heating, cooling, weighing) until two subsequent weights of the crucible agree within 0.005 g or less. After the final weighing add approximately 3.5 g of hydrated magnesium sulfate to the crucible and record the combined weight to the nearest 0.001 g.

Heat the crucible containing the magnesium sulfate in the same way the empty crucible was heated. Repeat heating, cooling, and weighing of the crucible containing the magnesium sulfate until subsequent weighings agree within 0.005 g.

Part C—Unknown

If your instructor assigns you an unknown, determine its water content by heating a weighed sample of it in a crucible. Use the procedure described in Part E for magnesium sulfate.

CALCULATIONS

Suppose a crucible that contained 3.583 g of an unknown hydrate lost 0.644 g of water. The percentage of water can be calculated as follows.

$$\% \text{ water } = \frac{\text{weight of water in sample}}{\text{original weight of sample}} \times 100\%$$

$$= \frac{0.644 \text{ g}}{3.533 \text{ g}} \times 100\% = 17.97\% = 18.0\% \text{ (to three significant figures)}$$

The number of moles of water of hydration per mole of hydrated compound may be calculated if the molecular weight of the anhydrous sample is known. Let us assume a molecular weight of 159.5 g/mole and calculate the moles of water of hydration in the previous example.

1. Calculate the number of moles of water.

$$\text{Moles of } H_2O = \frac{\text{weight of water}}{\text{molecular weight of water}} = \frac{0.644 \text{ g}}{18.0 \text{ g/mole}} = 3.58 \times 10^{-2} \text{ mole}$$

2. Calculate the number of moles of the anhydrous sample.

$$\text{Moles of sample} = \frac{\text{weight of anhydrous sample}}{\text{molecular weight of anhydrous sample}}$$

$$= \frac{3.583 \text{ g} - 0.644 \text{ g}}{159.5 \text{ g/mole}} = 1.843 \times 10^{-2} \text{ mole}$$

3. Determine the ratio of the number of moles of water to the number of moles of anhydrous sample.

$$\frac{\text{moles } H_2O}{\text{moles anhydrous compound}} = \frac{3.58 \times 10^{-2} \text{ mole}}{1.843 \times 10^{-2} \text{ mole}} = 1.94$$

4. Round to the nearest whole number.

$$\frac{\text{moles } H_2O}{\text{moles anhydrous compound}} = 2$$

Report Sheet

Name	Instructor/Section	Date

Part A—Classification of Hydration

	Original color before heating	H₂O appears?	Color of dry residue	Soluble in water?	Color of wet residue
Sucrose	_____	_____	_____	_____	_____
Copper sulfate	_____	_____	_____	_____	_____
Sodium sulfate	_____	_____	_____	_____	_____
Potassium chloride	_____	_____	_____	_____	_____
Cobalt chloride	_____	_____	_____	_____	_____

Part B—Water Content in Magnesium Sulfate

Weight of empty crucible

    After first heating (g)   (1a) _____

    After second heating (g)   (1b) _____

    After third heating (g)   (1c) _____

Weight of crucible and hydrated magnesium sulfate

    Before heating (g)   (2) _____

Weight of crucible and magnesium sulfate

    After first heating (g)   (3a) _____

    After second heating (g)   (3b) _____

    After third heating (g)   (3c) _____

Weight of anhydrous magnesium sulfate (g)   (4) _____

Weight of water lost (g)   (5) _____

Moles of water lost   (6) _____

Moles of water per mole of magnesium sulfate   (8) _____

Formula for the hydrated magnesium sulfate   (9) _____

Part C—Water Content in an Unknown Compound

Weight of empty crucible

    After first heating (g)   (1a) _____

    After second heating (g)   (1b) _____

    After third heating (g)   (1c) _____

Weight of crucible and unknown before heating (g)   (2) _____

Weight of crucible and unknown

    After first heating (g)   (3a) _____

    After second heating (g)   (3b) _____

    After third heating (g)   (3c) _____

Net weight of unknown after heating (g)   (4) _____

Weight of water lost (g)   (5) _____

Percent of water in unknown   (6) _____

Attach sheets with your calculations and the questions sheet.

QUESTIONS

1. Calculate the percent of water in each of the following hydrates.

    a. $Na_2B_4O_7 \cdot 10\ H_2O$

    b. $CoCl_2 \cdot 6\ H_2O$

    c. $CaCl_2 \cdot H_2O$

    d. $Ba(OH)_2 \cdot 8\ H_2O$

2. Calculate the number of grams of water that could be obtained from heating a 4.80 g sample of $Na_2SO_4 \cdot 10\ H_2O$.

3. Hydrated magnesium sulfate ($MgSO_4 \cdot x\ H_2O$) is heated until all the water is driven off. If 4.64 g of sample was heated to yield 2.27 g of anhydrous salt, what is the value for x?

4. Which of the compounds studied in Part A would you classify as hydrates? Explain your reasoning.

5. When sodium hydrogen carbonate ($NaHCO_3$) is heated, it decomposes yielding both carbon dioxide (CO_2) and water.

$$2\ NaHCO_3(s) \longrightarrow CO_2(g) + H_2O(g) + Na_2CO_3$$

Sodium carbonate (Na_2CO_3) is soluble in water. Describe an experiment that could show whether or not sodium hydrogen carbonate has water of hydration

EXPERIMENT
11   Synthesis of $Cu(NH_3)_4SO_4 \cdot H_2O$

APPARATUS

Büchner funnel, trap, aspirator, vials or bags for product, gummed labels, 16 × 150 mm test tubes, rubber policeman, stirring rod, filter paper.

REAGENTS

concentrated aqueous ammonia (NH_3), copper(II) sulfate pentahydrate ($CuSO_4 \cdot 5H_2O$), ethanol, acetone.

INTRODUCTION

When ammonia (NH_3) is added to an aqueous solution containing copper(II) ions, the deep blue complex cation, $Cu(NH_3)_4^{2+}(aq)$, is formed. Each of the four ammonia molecules are bonded to the central copper ion by a coordinate covalent bond.

In this experiment the salt $Cu(NH_3)_4SO_4 \cdot H_2O$ is isolated by adding alcohol to the aqueous solution of the complex salt. The alcohol is less polar than water and decreases the solubility of the salt.   Alcohol is less dense than water and although it is soluble in water, it can float on the top of water if the two liquids are not mixed. If the alcohol is poured carefully onto the surface of the liquid, the slow diffusion of the alcohol into the water results in the slow growth of the crystals. Slowly grown crystals of the salt $Cu(NH_3)_4SO_4 \cdot H_2O$ usually are long deep blue needles.

PROCEDURE

Prior to performing this experiment, study the following sections of this manual: A, "Handling Liquid Reagents" (pages 6-7); B, "Handling Solid Reagents" (pages 7-8); D, "Filtration" (pages 9-11); and N' "Use of the Balances" (pages 27-29).

In a 100 mL beaker dissolve with stirring 6.3 g of $CuSO_4 \cdot 5H_2O$ in 6 mL of water. In the hood, add 10 mL of concentrated aqueous ammonia and stir thoroughly. Note the color and compare it with the color of the $CuSO_4 \cdot 5H_2O$ prior to the addition of the ammonia.

The complex salt is precipitated by the addition of ethanol. **Carefully** pour 10 mL of ethanol down the side of the beaker so that the alcohol runs onto the top of the solution. **Carefully** set the beaker in your locker until the next laboratory session. Cover it with an inverted 400 mL beaker.

Avoid agitating the solution. The crystals of the complex salt will crystallize as long needles when the alcohol slowly diffuses into the aqueous solution. At the next session stir the mixture gently to avoid breaking the long crystals. Set up a suction filtration apparatus and prepare the following three wash solutions in 15 × 160 mm test tubes.

1. A mixture containing 5 mL of ethanol and 5 mL of concentrated aqueous ammonia.
2. 10 mL of ethanol.
3. 10 mL of acetone.

First wet the filter paper in the Büchner funnel with water. Turn on the faucet to start the suction. Use the solution in the beaker to transfer the bulk of the crystals into the funnel. Use a rubber policeman to scape the last of the crystals into the funnel. Quickly wash the crystals in the funnel with the ethanol-ammonia mixture. Then wash with pure ethanol and finally with the acetone. Add each wash solution quickly so that the crystals are not exposed to air until after the acetone has gone through the funnel.

Dry the crystals by blotting them between two pieces of filter paper. Then transfer the dry crystals quickly to a clean, dry previously weighed vial or plastic bag and weigh. Calculate the percent yield of $Cu(NH_3)_4SO_4 \cdot H_2O$.

$$Cu^{2+}(aq) + SO_4^{2-}(aq) + 4\,NH_3(aq) + H_2O \longrightarrow Cu(NH_3)_4SO_4 \cdot H_2O(s)$$

Tape a label that contains the following information onto the bag or vial.

Gross weight

Weight of vial

Net weight

Chemical formula

Date

Preparer's name

Turn in your sample to your instructor with your report. If a plastic bag was used it may be stapled directly to the report.

CALCULATIONS

Calculate the percent yield of your product. The mass of the product calculated from the mass of the reactant and the chemical equation is the theoretical yield. The actual yield is the mass of product [$Cu(NH_3)_4SO_4 \cdot H_2O$] actually obtained in the experiment. The percent yield is calculated

$$\frac{\text{Actual yield}}{\text{Theoretical yield}} \times 100\% = \text{percent yield}$$

Report Sheet

_____   _____ _____
Name   Instructor/Section   Date

Weight of $CuSO_4 \cdot 5H_2O$ and beaker (g)   (1) ————————

Weight of beaker (g)   (2) ————————

Weight of $CuSO_4 \cdot 5H_2O$ (g)   (3) ————————

Weight of $Cu(NH_3)_4SO_4 \cdot H_2O$ and container (g)   (4) ————————

Weight of container (g)   (5) ————————

Weight of $Cu(NH_3)_4SO_4 \cdot H_2O$ (g)   (6) ————————

Theoretical yield of $Cu(NH_3)_4SO_4 \cdot H_2O$ (g)   (7) ————————

% yield   (8) ————————

Attach sheet with your calculations. Answer the questions on the next page.

QUESTIONS

1. Compare the color of your product to that of the starting material CuSO$_4$•5H$_2$O.

2. Calculate the number of moles of NH$_3$ required to react with 3.9 g of CuSO$_4$•5H$_2$O to give a 100% yield of Cu(NH$_3$)$_4$SO$_4$•5H$_2$O.

3. Assuming that the concentration of concentrated aqueous NH$_3$ is 15 M, calculate the number of milliliters of ammonia required to react with 3.9 g of CuSO$_4$•5H$_2$O to give a 100% yield of Cu(NH$_3$)$_4$SO$_4$•5H$_2$O.

EXPERIMENT
12   Spectroscopy

APPARATUS

Flashlight, hydrogen and helium discharge tubes, spectroscope, power supply.

INTRODUCTION

The energy of an electron in an atom is quantized, or restricted to certain values. When an atom absorbs energy from a flame or electric discharge, it may absorb only the energy needed to raise an electron to a higher energy level. The atom is then said to be excited. When the electron returns to a lower energy level, the energy that had been absorbed is emitted as light energy. If the light has a wavelength between 400 and 700 nm, it is in the visible region. The light is emitted as an emission spectrum.

A spectroscope is used to study emission spectra. A schematic representation of a spectroscope is given in Figure 12-1. Light emitted from the excited atoms enters the instrument through the slit on the left. The slit allows only a narrow, well-defined band of light to pass. The prism breaks this band of light into its component wavelengths. If the light is white light, a rainbow with all wavelengths in the visible region appears on the screen. Because an excited atom emits light of only certain energies (wavelengths), a spectrum of lines results. These lines are of different colors, depending on their respective wavelengths. Since the spectrum of each element is different, spectra can be used identify substances. In this experiment you will study the spectra of hydrogen and helium.

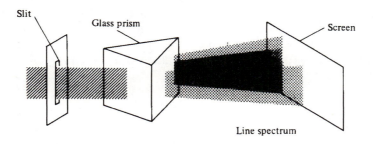

Figure 12-1. The line spectrum in a spectroscope.

The line spectrum obtained from an excited atom results from the emission of energy in only discrete amounts. These amounts, or quanta, of energy are known as photons. The energy of the photons associated with a given wavelength of light is given by the equation

$$E = \frac{h\,c}{\lambda} \qquad\qquad (1)$$

where

E = energy of photons in ergs
h = Planck's constant, 6.62×10^{-27} erg-sec
c = speed of light, 3.0×10^{10} cm/sec
λ ·= wavelength in cm

Thus, if the wavelength of a given line in an atomic spectrum is known, the difference in the energy levels between which the electrons fall can be calculated.

Each line of an atomic spectrum represents a different electronic transition. The wavelength of a given line in the spectrum of hydrogen is related to the electronic transition from which it resulted by the equation

$$\frac{1}{\lambda} = R \left[\frac{1}{n_x^2} - \frac{1}{n_y^2} \right] \qquad (2)$$

where

λ = wavelength of the spectral line in cm
R = Rydberg constant for hydrogen, $109{,}677.72$ cm^{-1}
n_x = lower energy level
n_y = lower energy level

Inserting the value of R into Equation (2) and rearranging gives the expression

$$\frac{1}{n_x^2} - \frac{1}{n_y^2} = \frac{9.12 \times 10^{-6} \text{ cm}}{\lambda} \qquad (3)$$

In this experiment you will determine the wavelengths for lines in the hydrogen spectrum. Each line corresponds to a different wavelength and thus to a different set of values for n_x and n_y. For a given wavelength, the principal quantum numbers n_x and n_y can be assigned from Equation (3). The value for $(9.12 \times 10^{-6} \text{ cm})/\lambda$ is calculated by inserting the value of a known wavelength in cm. Since both n_x and n_y are small integers, a table of values for the following expression can be prepared for various values of n_x and n_y.

$$\frac{1}{n_x^2} - \frac{1}{n_y^2}$$

The values from this table can be compared to the calculated values of $(9.12 \times 10^{-6} \text{ cm})/\lambda$. If a match, within experimental error, is found, the integers n_x and n_y are identified. If the wavelength of a spectral line is 1.216×10^{-6} cm, the value of $(9.12 \times 10^{-6} \text{ cm})/\lambda$ is 0.750. This value corresponds to $n_x = 1$ and $n_y = 2$ since, from the table,

$$\left[\frac{1}{n_x^2} - \frac{1}{n_y^2} \right] = \left[\frac{1}{1^2} - \frac{1}{2^2} \right] = \left[\frac{1}{1} - \frac{1}{4} \right] = 0.750$$

108

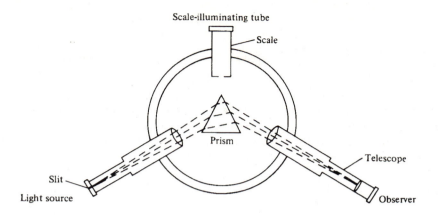

Figure 12-2 Components of the Spectroscope

This experiment is divided into three parts. In Part A the spectroscope is calibrated. Figure 12-2 is a schematic representation of the instrument that you will use. The light emitted by the excited atoms enters the slit on the left and passes through the prism, where the light is split into its component wavelengths. The resulting spectrum appears to be superimposed on an arbitrary scale that is viewed through the telescope. Calibrating the instrument consists of relating these scale divisions to wavelengths. Because the wavelengths of the lines of the helium spectrum are known, the helium spectrum will be used to calibrate the spectroscope. A plot of the wavelengths of the helium lines against the corresponding scale readings gives a graph that can be used to determine the wavelength of a line in any other spectrum when viewed with that instrument.

In Part B the spectroscope and the calibration curve from Part A are used to determine the wavelengths of the lines in the hydrogen spectrum. By using Equation (1), the energies of the photons associated with these lines will then be determined.

In Part C you will construct a partial energy-level diagram for the hydrogen atom. Your energy-level diagram should show the relative energy, to scale, of each level in the atom. Because you are investigating only the visible spectrum, only the transitions that emit visible light will be represented in your diagram. The construction of your energy-level diagram requires the knowledge that the lower energy level into which the electrons fall is the same for all transitions that produce light in the visible spectrum. You will find it convenient to use the lowest level as the base line of your diagram and then to represent the other energy levels as lines in their relative positions with respect to the base line.

PROCEDURE

Part A

Position the spectroscope so that the slit is directly in front of the helium discharge tube and quite close to it. For the spectroscope to operate effectively, the slit must be adjusted so that it is very nearly closed. If the slit is too wide, the beam will not be split effectively by the prism. Using the helium source, adjust the slit so that six or seven distinct lines are visible. This width should be used to view both the hydrogen and helium spectra.

Spectroscopy

After the instrument has been properly aligned and focused, observe the helium spectrum. Use a flashlight to illuminate the scale. In the data table, record to the nearest 0.1 scale division the scale reading next to the proper color for at least six lines. The first strong line should be red. A listing of the most intense helium emission lines according to colors and their relative intensities is provided in the Report Sheet. The violet line at 402.5 nm may not be visible to you. Additional low-intensity lines in the red and violet ranges might be visible to some individuals. Use only the most intense lines, as listed in the Report Sheet. Graph your data, using scale divisions as the abscissa (x-axis) and wavelengths (in centimeters) as the ordinate (y-axis). Review Section P, "Graphing," on pages 30-33 of this manual. Graph paper is found at the back of this manual.

Part B

Part B must be done with the same spectroscope used in Part A. Replace the helium discharge tube with the hydrogen discharge tube, and record the scale readings to the nearest 0.1 scale division for the lines of the hydrogen spectrum. An intense red line, an intense blue line, and a violet line should be visible.

Using the graph obtained in Part A to obtain the wavelengths of the three lines, calculate the energy in ergs of the photons in each line by Equation (1).

Part C

Calculate the values for $(1/n_x^2 - 1/n_y^2)$ for each set of principal quantum numbers n_x and n_y, in the first table in Part C of the Report Sheet. Calculate the value of $(9.12 \times 10^{-6}$ cm $/\lambda)$ for each line observed in the hydrogen spectrum. Using Equation (3), match your calculated value of $(1/n_x^2 - 1/n_y^2)$ and $(9.12 \times 10^{-6}$ cm $/\lambda)$ to determine the principal quantum numbers for the observed transitions for hydrogen.

Construct an energy-level diagram for the hydrogen atom, using the energies calculated in Part B. Use the lowest energy level for which you have data as the base line. Label your vertical axis clearly, and indicate the principal quantum number associated with each of the energy levels. Using vertical arrows, indicate the wavelength and energy associated with each electronic energy transition. Include the transition associated with the line at 410.1 nm.

One other line in the visible spectrum of hydrogen that probably cannot be seen with your instrument has a wavelength of 410.1 nm. Calculate the principal quantum numbers n_x and n_y for this transition also.

110

Report Sheet

Name	Instructor/Section	Date

Part A

Calibration of the spectroscope: helium spectrum

Line	Color	Relative Intensity	Wavelength (nm)	Wavelength (cm)	Scale Reading
1	Red	100	667.8	_____	_____
2	Yellow	1000	587.6	_____	_____
3	Light green	100	501.6	_____	_____
4	Dark green	50	492.2	_____	_____
5	Blue-green	40	471.3	_____	_____
6	Blue-violet	100	447.1	_____	_____
7	Violet	70	4026	_____	_____

The seventh line is not visible on some spectroscopes.

Part B

Hydrogen spectrum

Line	Color	Scale Reading	Wavelength (cm)	Energy of photons (ergs)
1	Red	_____	_____	_____
2	Blue	_____	_____	_____
3	Violet	_____	_____	_____
At 410.1 nm			4.10×10^{-5}	_____

Spectroscopy

Part C

Electronic transitions in hydrogen

n_x	n_y	$\dfrac{1}{n_x{}^2} - \dfrac{1}{n_y{}^2}$
1	2	0.750
1	3	_____
1	4	_____
1	5	_____
2	3	_____
2	4	_____
2	5	_____
2	6	_____
3	4	_____
3	5	_____
3	6	_____
3	7	_____

Wavelengths and quantum number assignments

Line	Color	Wavelength (cm)	$\dfrac{9.12 \times 10^{-6}\ \text{cm}}{\lambda}$	n_x	n_y
1	Red	_____	_____	_____	_____
2	Blue	_____	_____	_____	_____
3	Violet	_____	_____	_____	_____
At 410.1 nm		4.10×10^{-5}	_____	_____	_____

Attach sheets showing sample calculations, the questions sheet, and your graphs.

112

QUESTIONS

1. What should be the color of the line at 410.1 nm in the hydrogen spectrum?

2. What is the approximate wavelength of an orange line in a spectrum?

3. Describe the expected appearance of an emission spectrum of a gaseous mixture of hydrogen and helium.

4. The fit of a calibration curve depends on the number of points used in establishing a curve. Suppose additional lines of known wavelength at 624.6, 543.5, 498.0, 482.6, and 464.0 nm were available to you from other gases. Which two lines would be most effective for improving your calibration curve? Explain why.

5. What should be the wavelength of radiation for a transition from n = 2 to n = 1? Is this radiation in the visible light range?

EXPERIMENT

13   Charles's Law

APPARATUS

30 cc (mL) glass hypodermic syringe, 5 mm × 9 mm serum stopper, 600 mL beaker, ring stand, iron ring, clamp, wire gauze, buret clamp, Meker burner, matches, and rubber tubing.

INTRODUCTION

In 1787, Jacques Charles discovered that all gases expand when heated at constant pressure and that equal volumes of all gases expand to the same extent for the same increase in temperature. A mathematical statement of Charles's law would be

$$\text{Temperature} = m\,(\text{volume}) + b \tag{1}$$

In this experiment the temperature-volume relationship will be examined with a simple apparatus. A hypodermic syringe will be used to trap a definite volume of air. By placing the syringe in a water bath, the temperature of the trapped air can be controlled and measured. As the temperature of the bath rises, the plunger in the syringe will also rise. This indicates an increase in the volume of the trapped air with increasing temperature. The volume of the gas is obtained from the position of the plunger as read from the scale on the barrel of the syringe. The temperature is read from the thermometer in the bath.

PROCEDURE

Obtain from the storeroom a small serum stopper and a 30 cc glass syringe. Check the action of the plunger in the barrel. If the plunger does not move freely, wipe the plunger and the inside of the barrel with a clean towel. Then lubricate the plunger with graphite from a pencil. Two separate determinations will be run.

Part A

For the first determination pull out the plunger to about 20 cc and use the smaller end of the serum stopper to plug the hole in the barrel of the syringe (see Figure 13-1).

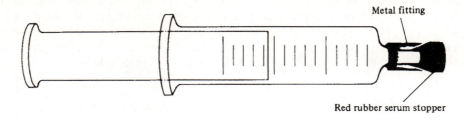

Figure 13-1. Glass syringe with serum stopper.

Attach the syringe assembly to a buret clamp and immerse it in an ice-water bath as shown in Figure 13-2. Do not squeeze the buret clamp on the barrel so tightly that the plunger drags against the barrel. Allow several minutes for the syringe to attain the same temperature as the ice water.

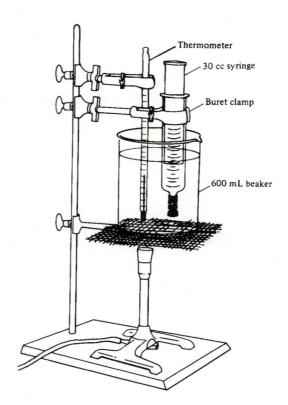

Figure 13-2. Apparatus for Charles's law determination.

Place a thermometer in the bath and read the temperature to the nearest 0.1°C. Gently rotate the plunger to overcome sticking against the barrel and read the volume to the nearest 0.1 cc. Record the temperature and volume on the Report Sheet. Always be sure that the volume of gas is maintained below the surface of the water bath.

Remove the ice water from the beaker, and add water at room temperature. After several minutes, take temperature and volume readings and record the data on the Report Sheet. Using a Meker burner, heat the water to approximately 50, 75, and 100°C. At each new temperature allow about 2 minutes for the syringe to attain the same temperature as the water in the bath. Record the volume of air for each temperature on the Report Sheet.

Part B

Repeat the entire procedure, starting this time with 10 cc of air at room temperature in the syringe instead of 20 cc. If, after dismantling the apparatus, water from the bath is discovered inside the syringe, the data must be discarded and the experiment redone with a dry syringe.

TREATMENT OF DATA

Make a graph of the temperature versus the volume of air. Read Section P, "Graphing" (pages 30-33), and use graph paper from the back of this manual. The temperature scale (ordinate) should read from –450 to 100°C, while the volume scale (abscissa) should read from 0 to 24 cc. Plot the data for both runs on the same graph. Use the symbol ⊗ to denote the points for the first determination (Part A) and the symbol ⊙ for the points of the second one (Part B).

For each determination draw the best straight line through the five points. Extrapolate each line to zero volume, and record the temperature on the Report Sheet. This value is b of Equation (1). Calculate the slope (m) of each line. Remember to use the correct units for the slopes and intercepts.

Report Sheet

_____ _____ _____
Name   Instructor/Section   Date

Part A

	Temperature (°C)	Volume (cc)
1.	_____	_____
2.	_____	_____
3.	_____	_____
4.	_____	_____
5.	_____	_____

Part B

	Temperature (°C)	Volume (cc)
1.	_____	_____
2.	_____	_____
3.	_____	_____
4.	_____	_____
5.	_____	_____

Analysis of Graphical Data

	Slope, m	Temperature at zero volume, b
Part A	_____	_____
Part B	_____	_____

Attach your graph and sample calculations. Answer the questions on the next page.

Charles's Law

QUESTIONS

1. Considering the accuracy of the volume measurement, which set of data in this experiment should give the most correct value of absolute zero?

2. How could additional measurements of the volume of the gas at other temperatures increase the accuracy of the determination of the temperature at which the volume of the gas is zero? Should temperatures lower than 0°C or greater than 100°C be selected? Why?

3. What effect, if any, would there be if the volume of the gas were determined at 15, 30, 40, 70, and 90°C instead of at the temperatures suggested in the procedure?

4. For each determination what is the temperature (t) at which the volume is zero? Are these values nearly identical or substantially different?

5. Assume that a new temperature scale (T) is devised for which $T = t - b$. Use the average of your two b values and substitute this expression for t in Equation (1). What relationship now exists between T and V?

6. At what temperature on the new scale is the volume of the air equal to zero?

7. Charles's law is a special case of the general gas law: $PV = nRT$. Rearrangement of this equation to a form comparable to Equation (1) yields $T = (P/nR)V$. The quantity (P/nR) corresponds to slope of the T versus V line. Explain why the slopes of the graphs for the two determinations are different.

EXPERIMENT

14   Molar Volume of Oxygen

APPARATUS

250 mL suction flask, 250 mL Erlenmeyer flask, pyrex test tube (25 × 200 mm), 250 mL beaker, 100 mL graduated cylinder, pinch clamp, two buret clamps, two ring stands, rubber tubing, glass tubing, thermometer, No. 4 one-hole rubber stopper, No. 6 one-hole rubber stopper.

REAGENTS

Potassium perchlorate ($KClO_4$) in individual vials (0.3 g), iron(III) oxide (Fe_2O_3).

INTRODUCTION

One mole of any gas occupies the same volume as 1 mole of any other gas if they are compared at the same conditions of temperature and pressure. In this experiment you will determine the volume of 1 mole of oxygen at standard temperature and pressure (STP).

You will use the decomposition of potassium perchlorate ($KClO_4$) in the presence of the catalyst iron(III) oxide (Fe_2O_3) as a source of oxygen.

$$KClO_4 \xrightarrow{Fe_2O_3} KCl + 2\ O_2$$

The loss of mass of the $KClO_4$–Fe_2O_3 mixture during the reaction is equal to the mass of oxygen evolved, and from this the number of moles of oxygen is calculated. The volume of oxygen will be determined by the displacement of water. From the volume of a specific mass of oxygen under known conditions of temperature and pressure, you will calculate the volume that would be occupied by 1 mole (32.0 g) of oxygen at standard temperature and pressure.

PROCEDURE

The preparation of the apparatus for this experiment requires cutting and fire polishing glass tubing. The glass tubing must be inserted into rubber stoppers.

Before doing any of these potentially hazardous operations, it is imperative that you study Section H, "Glassworking," on pages 16-17 of this manual.

You should also read following sections of this manual: E, "The Gas Burner" (pages 11-12); I, "Reading a Meniscus" (pages 17-18); M, "Use of the Laboratory Barometer" (pages 26-27); N, "Use of the Balances" (pages 27-29).

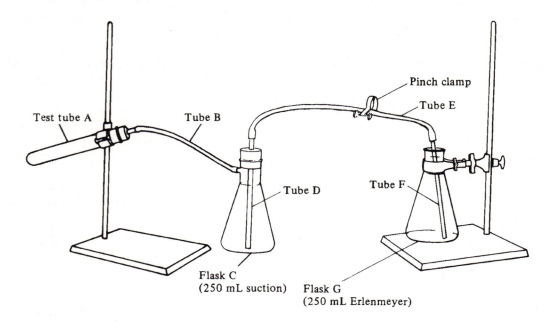

Figure 14-1. Experimental setup for molar volume of oxygen determination.

Assemble your apparatus as shown in Figure 14-1. Tube A should be a 25 × 200 mm hard glass test tube. Tube D should reach nearly to the bottom of flask C. All stoppers should fit tightly. Fill flask C almost to the neck with tap water and place about 50 mL of tap water in a beaker.

Note that no stopper is used in flask G. Placing a stopper in flask G would close the system and an explosion would occur upon heating.

Replace flask G with a beaker containing tap water. Remove the test tube from the apparatus. Obtain a vial containing about 0.3 g of potassium perchlorate from your instructor, and add all of the $KClO_4$ to the clean, dry test tube.

CAUTION: $KClO_4$ should not be discarded with paper, nor should it be weighed on paper. $KClO_4$ and paper can ignite spontaneously. Excess $KClO_4$ should be washed down the drain with large quantities of water.

Wipe any glycerine, used to insert the glass tube, from the stopper for tube A. Add 1 micro spatulaful of Fe_2O_3 (about 0.1 g) to the test tube. Determine the combined mass of the test tube and its contents to the nearest 0.0001 g on an analytical balance and record it on the Report Sheet. Tap the test tube to mix the contents.

Remove the pinch clamp from your apparatus. With the test tube removed, use a pipet bulb to blow on the free end of tube B until tubes D, E, and F are filled with water. By raising the beaker, adjust the water level in flask C until the level again is almost at the neck of the flask as shown in Figure 14-1. Attach and close the pinch clamp on tube E. Replace the beaker with a clean dry Erlenmeyer flask that is clamped in such a position that tube F reaches almost to the bottom of the flask.

Replace the test tube in the apparatus, placing the clamp near the top of the tube and seating the stopper snugly in the test tube. Remove the pinch clamp. If your system is airtight, no more than a drop or two of water should run out of tube F.

With the pinch clamp disconnected, gently heat the test tube and its contents until 75-100 mL of water has been displaced into flask G. Allow the apparatus to return to room temperature with the clamp disconnected. Some of the water that has been displaced into flask G will siphon back into flask C during the cooling process.

The pressures in flasks G and C are equalized by adjusting the height of flask G until the levels of the water in the two flasks are equal. Then reconnect the pinch clamp between the flasks. While being careful not to disturb the water remaining in tube F, remove flask G. Measure the water in flask G with a graduated cylinder. Record this value to the nearest 0.1 mL on your Report Sheet.

Disassemble the remaining apparatus and measure the temperature of the water in flask C, recording the reading as the temperature of the oxygen collected. Reweigh the cooled test tube and its contents on the same analytical balance and record the mass on the Report Sheet. Read and record the barometric pressure also.

TREATMENT OF DATA

The volume of oxygen is equal to the volume of water displaced. Since water has a significant vapor pressure, this pressure must be subtracted from the atmospheric pressure to yield the net pressure of oxygen gas in accordance with Dalton's Law of Partial Pressures. The vapor pressure of water at various temperatures is given in Table 14-1.

$$P_{atm} = P(O_2) + P(H_2O)$$

Table 14-1. Vapor pressure of water

Temperature (°C)	Pressure (mm Hg)	Temperature (°C)	Pressure (mm Hg)
17	14.5	24	22.4
18	15.5	25	23.8
19	16.5	26	25.2
20	17.5	27	26.7
21	18.6	28	28.4
22	19.8	29	30.0
23	21.1	31	33.7

Molar Volume of Oxygen

Using the pressure of oxygen, the temperature, and the volume, calculate the volume of oxygen gas at STP. The following combined gas law equation is useful.

$$\frac{P_1 V_1}{T_1} = \frac{P_2 V_2}{T_2}$$

Calculate the number of moles of oxygen in your sample. This is done by dividing the mass of oxygen, generated from the decomposition of potassium perchlorate, by the molecular weight of oxygen.

Calculate the ratio of volume/moles for your experiment. This ratio is the volume occupied by one mole of the gas at STP.

Report Sheet

_____ _____ _____
Name   Instructor/Section   Date

Initial mass of test tube and contents (g)   (1) _____

Final mass of test tube and contents (g)   (2) _____

Mass of oxygen evolved (g)   (3) _____

Volume of oxygen evolved (mL)   (4) _____

Temperature of oxygen (°C)   (5) _____

Temperature of oxygen (K)   (6) _____

Barometric pressure (mm Hg)   (7) _____

Vapor pressure of water at above temperature (mm Hg)   (8) _____

Pressure of oxygen (mm Hg)   (9) _____

Volume of oxygen at STP (mL)   (10) _____

Molar volume of oxygen   (11) _____

Percent error (accepted value = 22.392 L)   (12) _____

Attach sample calculations. Answer the questions on the next page.

QUESTIONS

1. Suppose you had used triple the amount of Fe_2O_3. How would this have affected your calculated molar volume of oxygen?

2. During the experiment you probably did not heat the test tube long enough to decompose all of the $KClO_4$. Why is complete decomposition of $KClO_4$ unnecessary?

3. Why must tube F reach almost to the bottom of flask G during the cooling period?

4. What problem would be introduced if you did not equalize the water levels in the reservoir (flask C) and receiving vessel (flask G) before closing the pinch clamp?

5. The gas contained in flask C consists of oxygen and water vapor. How is the pressure of oxygen alone determined?

6. Suppose, after filling tubes D and F with water and closing pinch clamp E, you had spilled some of the water in tube F on the bench top. What effect would this have on your calculated molar volume of oxygen? Explain.

EXPERIMENT
15 Analysis of a Mixture

APPARATUS

Crucible (size 0) with cover, clay triangle, ring stand, iron ring, clamp, Meker burner, rubber tubing, matches, microspatula.

REAGENTS

Approximately 0.5 g $KClO_3$ / KCl mixture, Fe_2O_3.

INTRODUCTION

The composition of a mixture can be determined by either physical or chemical methods. One example of a physical method was illustrated in Experiment 4 where the difference in the solubility of two compounds was used. In a chemical method, a chemical reaction is used to determine the amount of one component of a mixture which reacts while the other component remains unchanged. Based on known stoichiometric relationships, the change in mass of the mixture is used to calculate the amount of the original reactive component present in the mixture. The mass of the unreactive component is determined by difference.

In this experiment, the composition of a mixture of potassium chlorate ($KClO_3$) and potassium chloride is determined. Potassium chlorate decomposes to yield potassium chloride and oxygen gas. The reaction is catalyzed by iron(III) oxide. The product KCl as well as the original KCl present in the mixture undergo no reaction when heated.

$$2\ KClO_3 \xrightarrow{\ Fe_2O_3\ } 2\ KCl\ +\ 3\ O_2$$

The loss of mass of the $KClO_3$ during the reaction corresponds to the mass of oxygen evolved. This quantity is determined by weighing a reaction vessel containing the mixture prior to heating and also after completion of the reaction. Using stoichiometric relationships derived from the balanced equation, the amount of $KClO_3$ originally present in the reaction mixture is determined.

Analysis of a Mixture

PROCEDURE

Study Section N, "Use of the Balances," on pages 27-29 of this manual. To avoid errors you should use the same analytical balance for all weighings in this experiment. Study Sections E, "The Gas Burner" (pages 11-12), and G, "Use of the Crucible" (pages 14-15).

CAUTION: The hot crucible as well as the clay triangle and ring can cause severe burns to the skin. Immediately place your hand under running water if you inadvertently touch any hot portion of the apparatus, and call your instructor.

Obtain a porcelain crucible with a cover and wash both with soap and water. After drying with a towel, place the porcelain crucible on a clay triangle supported by an iron ring (see Figure 15 on page 15 of this manual). Adjust the gas burner to a cool flame and heat the crucible gently for a few minutes. Then adjust the burner to give a very hot flame. Open the gas jet all the way and then adjust the mixture control at the bottom of the burner until a blue cone appears and a hissing sound is heard. Place the crucible in the upper part of the blue cone. A properly adjusted burner will make the crucible glow orange-red after only a few minutes.

After heating the glowing crucible for 10 minutes, turn off the burner. From this point on, handle the crucible and cover only with tongs, to prevent the moisture of your hands from being deposited on them. Using the tongs, place the covered crucible onto a ceramic mat (Figure 17 on page 15). During cooling, the lid should be placed tightly on the crucible. After the crucible and lid have cooled to room temperature, determine their mass with the analytical balance. Record this mass on the Report Sheet.

Reheat the crucible with the cover ajar for 5 minutes. After allowing time for the apparatus to cool to room temperature, weigh the crucible and cover on the analytical balance. If the second weight differs from the first weight by more than 0.005 g, repeat the heating process until subsequent weighings agree within 0.005 g.

Using the analytical balance, weigh from 0.4 to 0.5 g of a mixture (obtained from your instructor) into the crucible. Replace the cover and reweigh on the analytical balance. Remove the cover and add about 50 mg of iron(III) oxide. Replace the cover and reweigh on the analytical balance. Record all masses on your Report Sheet.

Use a moderate flame to slowly heat the crucible. The decomposition reaction may initially be vigorous, and you should heat carefully to avoid ejecting some of the chlorate mixture from the crucible. When the reaction has subsided, place the cover, slightly ajar, on the crucible and heat strongly for 5 minutes. Allow the crucible, cover, and contents to cool to room temperature. Weigh the crucible, cover, and contents on the analytical balance. Record the mass on your Report Sheet.

If sufficient time remains in the laboratory period, clean and dry the crucible and repeat the experiment. If two trials are done, calculate the average % $KClO_3$ in the reaction mixture.

CALCULATIONS

A student obtained the following data for the determination of the composition of the mixture of potassium salts.

Mass of crucible and lid	= 22.9230 g
Mass of crucible, lid, and mixture	= 23.4063 g
Mass of crucible, lid, and mixture and iron(III) oxide	= 24.4102 g
Mass of system after heating	= 24.3290 g

1. Calculate the mass of mixture used.

    Mass of mixture = (mass of mixture + crucible + lid) – (mass of crucible + lid)

    = 23.4063 g – 22.9230 g

    = 0.4833 g

2. Calculate the mass of oxygen gas evolved.

    Mass of oxygen = (mass of mixture + Fe_2O_3 + crucible + lid) – (mass after heating)

    = 24.4102 g – 24.3290 g

    = 0.0812 g

3. Calculate the moles of oxygen gas evolved.

    Moles of O_2 = $\dfrac{\text{mass of oxygen}}{\text{molecule weight of } O_2}$

    = $\dfrac{0.0812 \text{ g}}{32.0 \text{ g/mole}}$

    = 2.54×10^{-3} mole O_2

4. Calculate the moles of $KClO_3$ in the mixture based on the balanced equation.

    Moles of $KClO_3$ = $\dfrac{2 \text{ moles } KClO_3}{3 \text{ moles of } O_2}$ × $(2.54 \times 10^{-3}$ mole $O_2) = 1.69 \times 10^{-3}$ mole $KClO_3$

5. Calculate the mass of $KClO_3$ in the mixture.

$$\text{Mass of } KClO_3 = \frac{122.55 \text{ g } KClO_3}{1 \text{ mole of } KClO_3} \times (1.69 \times 10^{-3} \text{ mole } KClO_3)$$

$$= 0.207 \text{ g } KClO_3$$

6. Calculate the percent $KClO_3$ in the reaction mixture.

$$\% \, KClO_3 = \frac{\text{Mass of } KClO_3}{\text{Mass of mixture}} \times 100 \, \%$$

$$= \frac{0.207 \text{ g } KClO_3}{0.4833 \text{ g mixture}} \times 100 \, \%$$

$$= 42.8\% \, KClO_3$$

Report Sheet

Name   Instructor/Section   Date

	Trial I	Trial II
Mass of crucible and cover (g)	(1a) _____	(1a) _____
	(1b) _____	(1b) _____
Mass of crucible, cover, and mixture (g)	(2) _____	(2) _____
Mass of crucible, cover, mixture and oxide (g)	(3) _____	(3) _____
Mass of system after heating (g)	(4a) _____	(4a) _____
	(4b) _____	(4b) _____
Mass of oxygen (g)	(5) _____	(5) _____
Moles of oxygen molecules	(6) _____	(6) _____
Moles of $KClO_3$ in mixture	(7) _____	(7) _____
Mass of $KClO_3$ in mixture (g)	(8) _____	(8) _____
Percent $KClO_3$ in mixture	(9) _____	(9) _____
Average Percent $KClO_3$ in mixture	(10) _____	

Attach sheets showing your calculations and the questions sheet.

QUESTIONS

1. How would your calculated % $KClO_3$ compare to the true value if you failed to heat the crucible sufficiently to completely decompose the compound?

2. How would your calculated % $KClO_3$ compare to the true value if some of the mixture were ejected from the crucible during heating?

3. How would your calculated % $KClO_3$ compare to the true value if the crucible were not dry prior to adding the mixture to the empty crucible?

4. What effect would the use of twice the amount of recommended iron(III) oxide have on the calculated % $KClO_3$ compared to the true value?

5. How would the accuracy of the calculated % $KClO_3$ be affected if half of the suggested quantity of the mixture were used?

6. How many moles of Fe_2O_3 were used in Trial I of this procedure?

EXPERIMENT

16 Determination of Zinc Ion by Titration

APPARATUS

25 mL buret, 5 mL and 10 mL pipets, 125 mL Erlenmeyer flasks, microspatula, ring stand, two buret clamps.

REAGENTS

Standardized EDTA solution close to 0.01 M, buffer solution (0.5 M NH_3, 0.5 M NH_4Cl adjusted to pH 10), Eriochrome Black T–potassium chloride mixture, unknown Zn^{2+}(aq) samples.

INTRODUCTION

The determination of the concentration of a metal ion in a solution is an important analytical technique. Examples of such analyses include the determination of calcium ion in hard water and the measurement of materials discharged from the effluent of manufacturing plants.

In this experiment the presence of Zn^{2+} will be determined with ethylenediaminetetraacetic acid (EDTA). This complex organic compound can exist as an anion with –4 charge and will be symbolized E^{4-}. The anion reacts with many metal ions in a one to one ratio to form a complex ion. For zinc the complex is represented as ZnE^{2-}.

$$Zn^{2+} + E^{4-} \rightleftharpoons ZnE^{2-}$$

It is necessary to determine when exactly one E^{4-} per one Zn^{2+} is present. This determination is done by titrating (gradually adding) E^{4-} to a solution of Zn^{2+} in the presence of an indicator.

The indicator used in this experiment is another complex organic molecule called Eriochrome Black T. In solution the indicator exists as an anion represented by T^-. In the presence of zinc ion the indicator forms a one to one complex ion.

$$Zn^{2+} + T^- \rightleftharpoons ZnT^+$$
$$\text{(pink)}$$

The complex ion has a pink color. Only a small amount of T^- is used to give enough ZnT^+ to see the color. All of the T^- is complexed to the zinc ion. Since the Zn^{2+} is present in excess, the majority of the zinc ion is not complexed. As EDTA is added to a solution during the titration, the EDTA can react with both the zinc ion and the colored complex ion, ZnT^+, as follows.

$$Zn^{2+} \; + \; E^{4-} \; \longrightarrow \; ZnE^{2-}$$
(colorless) (colorless)   (colorless)

$$ZnT^+ \; + \; E^{4-} \; \longrightarrow \; ZnE^{2-} \; + \; T^-$$
(pink)   (colorless)   (blue)

Only a small amount of indicator is used in the reaction. Thus, during the titration (careful addition of EDTA solution), the E^{4-} reacts with the free Zn^{2+} to give colorless ZnE^{2-}. At the end of the titration, the E^{4-} finally reacts with the ZnT^+ to replace T^- with E^{4-}. The T^- released is blue, so a color change from pink to blue results.

According to the balanced equation, the number of moles of E^{4-} used equals the number of moles of Zn^{2+} present in the sample. The number of moles of E^{4-} used is calculated as

$$(\text{Liters of } E^{4-})\,(\text{molarity of } E^{4-}) \; = \; \text{moles of } E^{4-}$$

$$\text{Moles of } E^{4-} \; = \; \text{moles of } Zn^{2+}$$

$$\text{Concentration of } Zn^{2+} \; = \; \frac{\text{moles of } Zn^{2+}}{\text{volume of sample}}$$

PROCEDURE

Before starting this experiment you should study Sections I, "Reading a Meniscus,' (pages 17-18), and J, "Volumetric Analysis" (pages 18-22), in this manual.

Obtain a bottle of an unknown Zn^{2+} solution from the instructor. Obtain approximately 40 mL of standardized EDTA solution to rinse out and fill your 25 mL buret. Mark the exact concentration (to four significant figures) on the Report Sheet.

Carefully pipet 5.00 mL of your unknown solution of Zn^{2+} into a clean 125 mL Erlenmeyer flask and add approximately 10 mL of distilled water from a graduated cylinder. Then add approximately 5 mL of a solution called a buffer. Obtain the Eriochrome Black T mixture from the instructor and add about one-third of a microspatula of the indicator to the flask. A light pink color should appear as the indicator slowly dissolves. Return the indicator to the instructor.

Titrate the Zn^{2+} with EDTA of known concentration. (The exact concentration, which will be close to 0.01 M, should be copied from the bottle.) Gently swirl the contents of the flask during the titration until sufficient EDTA has been added to fade the pink color completely. Record the volume of EDTA used. Addition of one drop past the endpoint will change the color to blue. Read and record the volume to the nearest 0.01 mL.

Repeat the titration procedure two more times. However, choose a 10.00 mL sample of Zn^{2+} if the amount of EDTA used in your first titration was less than half the capacity of the buret. If 10.00 mL is used, add 20 mL of distilled water and 10 mL of the buffer. Obtain a sufficient quantity of standardized EDTA for two additional trials. Repeat the trials using the same 5.00 or 10.00 mL volume of Zn^{2+} unknown until two subsequent volumes agree to within 0.10 mL.

Calculate the molarity of Zn^{2+} ions in each of your three trials. If all three values determined are of comparable accuracy, average them to obtain the best experimental average value.

Report Sheet

		Name		Instructor/Section	Date

		Trial 1	Trial 2	Trial 3
Molarity of EDTA	(1)	_____	_____	_____
Initial buret reading	(2)	_____	_____	_____
Final buret reading	(3)	_____	_____	_____
Volume of EDTA delivered	(4)	_____	_____	_____
Moles of EDTA delivered	(5)	_____	_____	_____
Moles of Zn^{2+}	(6)	_____	_____	_____
Volume of Zn^{2+} taken	(7)	_____	_____	_____
Molarity of Zn^{2+}	(8)	_____	_____	_____
Average molarity of Zn^{2+}		_____		

Attach sheets with sample calculations. Answer the questions on the next page.

Determination of Zinc Ion by Titration

QUESTIONS

1. An air bubble is present in the stopcock at the start of the titration but is passed out during the titration. Would this error make your calculated molarity of zinc ions higher or lower than the true value? Explain your answer clearly.

2. How would your results be affected if a drop of EDTA on the tip of the buret were released after the end point? Explain your answer clearly.

3. How would your results be affected if 10 mL of tap water containing Fe^{3+} instead of 10 mL of distilled water were used to prepare the test solutions for this experiment?

4. Devise a procedure by which you could test your assumption that EDTA and Zn^{2+} form a one-to-one complex ion.

5. If the stoichiometry of the reaction were two Zn^{2+} per EDTA, what volume of EDTA would have been used in your experiment?

17 Analysis of a Bleach Solution

APPARATUS

25 mL buret, 125 mL Erlenmeyer flasks, 50 mL beakers, 10 × 75 mm test tubes, eye dropper, microspatula, ring stand, buret clamp.

REAGENTS

Commercial bleach or "artificial bleach" unknown solution, standardized 0.0500 M $Na_2S_2O_3$ solution, solid KI, 6 M HCl, soluble starch indicator ("Vitex").

INTRODUCTION

The active ingredient in commercial bleach is sodium hypochlorite (NaOCl). The hypochlorite ion is an oxidizing agent and is converted to chloride ion in a redox reaction. The net change in oxidation state is 2, because the oxidation numbers of chlorine in OCl^- and Cl^- are +1 and –1, respectively.

The effect of bleach in cleaning clothes is its ability to oxidize colored material to colorless products. The clothes aren't any cleaner, but they look whiter. In this experiment you will determine the weight percent concentration of sodium hypochlorite in bleach. A typical value is 5% sodium hypochlorite by weight. You will determine the number of moles of the hypochlorite ion in a selected mass of bleach and convert those quantities into weight percent.

The hypochlorite ion will oxidize iodide ion to form iodine, which in the presence of excess iodide ion yields the tri-iodide ion, I_3^-. The balanced net ionic equation is

$$OCl^- + 3\,I^- + H_2O \longrightarrow I_3^- + Cl^- + 2\,OH^- \tag{1}$$

The amount of tri-iodide ion formed depends on the number of moles of hypochlorite ion in the sample of bleach. The amount of tri-iodide ion formed can be determined by titration with standardized sodium thiosulfate solution. The net ionic equation for the reaction of thiosulfate ($S_2O_3^{2-}$) with tri-iodide ion is

$$I_3^- + 2\,S_2O_3^{2-} \longrightarrow 3\,I^- + S_4O_6^{2-} \tag{2}$$

By combining equations (1) and (2) the net chemical equation relating the amount of hypochlorite to the amount of thiosulfate used to titrate the tri-iodide formed is

$$OCl^- + 2\,S_2O_3^{2-} + H_2O \longrightarrow S_4O_6^{2-} + Cl^- + 2\,OH^- \tag{3}$$

The end point of the titration of tri-iodide ion by thiosulfate is determined by a color change from the dark blue of a complex of starch and iodine to a colorless solution. Unlike other indicators, the necessary starch is not added until near the end point. The color of tri-iodide solutions are light yellow. This color fades as the added thiosulfate reduces the tri-iodide ion. At the point when the yellow color has almost disappeared, the starch is added to form an intense blue colored starch-iodine complex. The titration then is continued to eliminate this blue color.

PROCEDURE

Before performing this procedure, study the sections "Reading a Meniscus" (pages 17-18) and "Volumetric Analysis" (pages 18-22).

Obtain approximately 3 mL of bleach in a 10 × 75 mL test tube. Determine the mass of a 50 mL beaker. Using an medicine dropper, add approximately 1 g of the bleach to the beaker and determine the mass of the sample and beaker using the analytical balance. Using a 10 mL graduated cylinder, add approximately 10 mL of distilled water to the sample. Quantitatively transfer the dilute solution to a 125 mL Erlenmeyer flask by rinsing the contents of the beaker into the flask using several 5 mL portions of distilled water.

> **CAUTION: Bleach can cause chemical burns and damage your clothes. It is also an eye irritant. If you spill the bleach, wash the contaminated area and keep your hands away from your eyes. Report the spill to your instructor.**

Add 2.0 g of sodium iodide to the flask and gently swirl the resulting solution to generate the tri-iodide ion. Add 5 mL of 6 M HCl using a 10 mL graduated cylinder. Note that the HCl **must not** be added prior to the addition of the sodium iodide.

> **CAUTION: The 6 M HCl can cause severe skin burns. Immediate washing of the affected area and prompt reporting to the instructor are important to avoid skin burns if the hydrochloric acid is spilled.**

Obtain approximately 40 mL of standardized 0.05 M sodium thiosulfate solution to rinse out and fill your 25 mL buret. Mark the exact concentration of the solution on your Report Sheet. Titrate the triodide ion using the sodium thiosulfate solution until the yellow color has nearly faded. Add half a microspatula of solubilized starch and slowly resume the titration. The color change from dark blue to colorless occurs suddenly. Record the volume of standardized thiosulfate used on your Report Sheet.

Repeat the analysis of the bleach solution using a second sample. If sufficient time is available, a third analysis should be done.

Report Sheet

		Name		Instructor/Section		Date

		Trial I	Trial II	Trial III
Mass of beaker and bleach solution (g)	(1)	_____	_____	_____
Mass of beaker (g)	(2)	_____	_____	_____
Mass of bleach solution (g)	(3)	_____	_____	_____
Molarity of $Na_2S_2O_3$	(4)	_____	_____	_____
Initial buret reading	(5)	_____	_____	_____
Final buret reading	(6)	_____	_____	_____
Volume of $Na_2S_2O_3$ delivered	(7)	_____	_____	_____
Moles of $Na_2S_2O_3$ delivered	(8)	_____	_____	_____
Moles of OCl^- in bleach sample	(9)	_____	_____	_____
Mass of NaOCl in bleach sample	(10)	_____	_____	_____
Weight percent of NaOCl	(11)	_____	_____	_____
Average weight percent	(12)	_____		

Attach your calculations and the answers to the questions on the following page.

Analysis of a Bleach Solution

QUESTIONS

1. Does the quantity of water used to wash the dilute bleach solution into the Erlenmeyer flask affect the calculation of the weight percent of the bleach solution?

2. Assume that an air bubble is present in the stopcock at the start of the titration and is passed out during the titration. Would this error make your calculated weight percent of the bleach solution higher or lower than the true value? Explain your answer clearly.

3. How would your results be affected if a drop of sodium thiosulfate on the tip of the buret were released after the end point? Explain your answer clearly.

4. How would the results of your experiment be affected if a small amount of a reducing agent were present in the beaker used to weigh the bleach solution?

5. The reason why the HCl solution is added only after the iodide ion is added to the flask is that chloride ion reacts with hypochlorite ion to form chlorine gas. Write a net ionic equation for this reaction.

6. Assuming that your sample of bleach has a density of 1.02 g/mL, calculate its molar concentration.

EXPERIMENT

18 Equilibrium

APPARATUS

Six 16 × 150 mm test tubes, test tube rack, spectrophotometer, cuvettes, and tissue paper for cleaning cuvettes.

REAGENTS

4.0×10^{-4} M sodium thiocyanate solution (NaSCN), 0.20 M, 0.040 M, 0.020 M, 0.010 M, and 0.0050 M solutions of iron(III) nitrate [$Fe(NO_3)_3$].

INTRODUCTION

In many chemical reactions the conversion of reactants to products is incomplete. No matter how long the reaction is allowed to continue, a quantity of each reactant always remains. In such cases an equilibrium has been established between the reactants and the products.

Consider the general equilibrium expression

$$a A + b B \rightleftharpoons c C + d D \tag{1}$$

where a, b, c, and d are the coefficients of the balanced equation. At the start of the reaction, A and B react to form C and D. As the concentrations of A and B decrease, the forward reaction rate slows down. As the concentrations of C and D increase, the reaction rate of C and D to produce A and B increases. Eventually the rate of the forward reaction equals the rate of the reverse reaction, and the system is at equilibrium.

At equilibrium, the concentrations of the reacting species in the general expression (1) are governed by the mathematical equation

$$K = \frac{[C]^c [D]^d}{[A]^a [B]^b} \tag{2}$$

where K is the equilibrium constant and the square brackets refer to concentrations in moles per liter. Again a, b, c, and d are the coefficients of the balanced equation (1). K is constant for a given temperature and is independent of the initial concentrations of the reacting species.

Equilibrium

In this experiment you will study the equilibrium of the iron(III) ion and the thiocyanate ion with the $FeSCN^{2+}$ complex ion.

$$Fe^{3+} + SCN^- \rightleftharpoons FeSCN^{2+} \tag{3}$$

The equilibrium constant expression for this reaction is

$$K = \frac{[FeSCN^{2+}]}{[Fe^{3+}][SCN^-]} \tag{4}$$

You are to determine experimentally the equilibrium constant for the reaction.

To show that the equilibrium constant is independent of the initial concentrations of the reactants, you will mix solutions of varying concentrations of the Fe^{3+} and SCN^- ions. You will then determine the equilibrium concentrations of the three species in each mixture and from these data calculate the equilibrium constant.

Since the complex ion $FeSCN^{2+}$ is blood-red in color, its equilibrium concentration can be determined with a spectrophotometer. The intensity of the color is directly proportional to the concentration of the colored species. This intensity is measured by absorbance (A) as given in Equation (5).

$$\text{Intensity of color} = A = k\,[FeSCN^{2+}] \tag{5}$$

In this experiment you will prepare a standard solution of known concentration and determine its absorbance. You then can calculate the proportionality constant k. Then the concentration of any unknown solution can be calculated from its measured absorbance.

To prepare a standard solution of $FeSCN^{2+}$ ion, you will use an excess of Fe^{3+} with a limited amount of SCN^- ion. Under these conditions, assume that there is so much Fe^{3+} present that all the SCN^- has been converted to $FeSCN^{2+}$.

In the standard solution, the concentration of the $FeSCN^{2+}$ ion is equal to the initial concentration of SCN^- after mixing with the Fe^{3+} but before a reaction occurs. This initial concentration of SCN^- would be less than the labeled concentration of the NaSCN stock solution, because mixing with another solution results in a dilution. The concentration after dilution can be calculated from the dilution formula.

$$M_1 V_1 = M_2 V_2 \tag{6}$$

M_1 and V_1 refer to the molarity and the volume of a species such as SCN^- before mixing, whereas M_2 and V_2 refer to the molarity and volume after mixing but before any reaction occurs. Note that mixing two solutions dilutes both of them.

Suppose that 20 mL of a 0.30 M solution of substance X is mixed with 10 mL of a 0.12 M solution of substance Y to give 30 mL of solution. The initial molarity of substance X after mixing but before reaction can be calculated from

$$M_1 V_1 = M_2 V_2$$
$$(0.30\ M)(20\ \text{mL}) = M_2\ (30\ \text{mL})$$
$$M_2 = 0.20\ M$$

The initial molarity of substance Y after mixing but before reacting is calculated in the same way.

$$M_1 V_1 = M_2 V_2$$
$$(0.12\ M)(10\ \text{mL}) = M_2\ (30\ \text{mL})$$
$$M_2 = 0.040\ M$$

The initial concentrations of Fe^{3+} and SCN^- in this experiment will be calculated after mixing the two solutions. Consider again the reaction

$$Fe^{3+} + SCN^- \rightleftharpoons FeSCN^{2+}$$

and note that, for each mole of $FeSCN^{2+}$ formed, 1 mole of Fe^{3+} reacts with 1 mole of SCN^-. Thus the concentration of Fe^{3+} is decreased as a result of the reaction that converts Fe^{3+} into the complex ion. At equilibrium the relationship is given by

$$[Fe^{3+}]_{equil} = [Fe^{3+}]_{init} - [FeSCN^{2+}]_{equil} \tag{7}$$

Similarly, the concentration of SCN^- is decreased as a result of its conversion into the complex ion. At equilibrium the relationship is given by

$$[SCN^-]_{equil} = [SCN^-_{init}] - [FeSCN^{2+}]_{equil} \tag{8}$$

PROCEDURE

Study the instructions given in Section L, "Spectrophotometry," on pages 24-25 of this manual for the proper use of the spectrophotometer. The instrument contains expensive electronic and optical components that must be handled carefully to prevent damage.

Do not make any adjustments that you do not understand.

In this experiment, the spectrophotometer should be set at 450 nanometers. At this wavelength the absorbance due to $FeSCN^{2+}$ is the strongest and thus the results will be more accurate.

Prepare five 16 × 150 mm test tubes for this experiment by washing with distilled water and drying. Number the tubes 1 through 5 and pipet 5.00 mL of $4.0 \times 10^{-4}\ M$ NaSCN to each tube. Into tube 1 pipet 5.00 mL of 0.20 M $Fe(NO_3)_3$ and use the resultant solution as your standard solution of $FeSCN^{2+}$ ion. Calculate the concentration of $FeSCN^{2+}$ in this tube using Equation (6).

Into tubes 2 through 5, pipet 5.00 mL of the $Fe(NO_3)_3$ solution indicated below.

Tube	$Fe(NO_3)_3$ solution to be used (M)
2	0.040
3	0.020
4	0.010
5	0.0050

Obtain two cuvettes from your instructor. Wash and rinse both of them. Add distilled water to one cuvette and set it aside as your reference cell. The other cuvette is your sample cell. Insert the cuvette with distilled water and set the %T to 100%. Remove the cuvette and insert the sample cuvette containing the contents of tube 1. Record the absorbance on the Report Sheet.

Rinse the sample cell two or three times with small amounts of each new solution before filling the tube for the absorbance determination. This procedure ensures that the concentration of the solution in the cuvette is the same as in the sample solution and is not contaminated by solutions previously placed in the cuvette.

Record the absorbance of each solution on your Report Sheet. Calculate the concentrations of Fe^{3+} and SCN^- after mixing but before any reaction has occurred.

Assuming that the SCN^- in tube 1 is completely converted to $FeSCN^{2+}$, use the absorbance to calculate the proportionality constant k according to Equation (5). Using this proportionality constant, and the absorbances of the solutions in tubes 2 through 5, calculate the equilibrium concentration of $FeSCN^{2+}$ in each tube. Then calculate the equilibrium concentrations of Fe^{3+} and SCN^- from Equations (7) and (8).

Using the equilibrium constant expression, calculate K for the solutions in tubes 2 through 5. Determine the average equilibrium constant from your data. Indicate if you discard any data as suspect. (You may redetermine any value that you wish by repeating that part of the experiment.)

Report Sheet

Complete the following data table.

Tube	Initial concentration (after mixing, before reaction)		Absorbance at 450 nm
	$[Fe^{3+}]_{init.}$	$[SCN^-]_{init.}$	
1	_____	_____	_____
2	_____	_____	_____
3	_____	_____	_____
4	_____	_____	_____
5	_____	_____	_____

Concentration of $FeSCN^{2+}$ at equilibrium in tube 1   _____

Value of proportionality constant k   _____

Using Equation (5), calculate the equilibrium concentrations of $FeSCN^{2+}$ in tubes 2 through 5 from the absorbance for each tube. Obtain the equilibrium concentrations of Fe^{3+} and SCN^- by subtracting the concentration of $FeSCN^{2+}$ at equilibrium from the initial concentrations of Fe^{3+} and SCN^-, as in Equations (7) and (8).

Tube	$[FeSCN^{2+}]_{equil.}$	$[Fe^{3+}]_{equil.}$	$[SCN^-]_{equil.}$
2	_____	_____	_____
3	_____	_____	_____
4	_____	_____	_____
5	_____	_____	_____

    Calculate the equilibrium constant K for each of the chemical systems in tubes 2 through 5 by substituting the equilibrium concentrations into the equilibrium expression (4).

	Tube 2	Tube 3	Tube 4	Tube 5
K	_____	_____	_____	_____
$K_{average}$	_____			

Attach sheets showing sample calculations and answers to the questions sheet.

QUESTIONS

1. How would the absorbance of the standard solution (tube 1) have been affected if 0.30 M $Fe(NO_3)_3$ had been used in place of 0.20 M? Remember that for the standard solution we have assumed that the reaction has been driven completely to $FeSCN^{2+}$. If it changes, calculate the new absorbance based on your values.

2. How would the absorbance of the standard solution have been affected if 0.00050 M NaSCN had been used in place of 0.00040 M? If it changes, calculate the new absorbance based on your values.

3. What would be the effect of improper rinsing of a cuvette prior to introducing a new solution? Considering the order of determination, how would the actual concentration of the $FeSCN^{2+}$ in the cuvette differ from your sample in the test tube?

4   The concentrations of Fe^{3+} and SCN^- are calculated by difference. In which tube is the concentration of Fe^{3+} least accurately known?

5. Assume that it is necessary to determine the $[Fe^{3+}]$ in a solution that may be in the 0.0001 to 0.0005 M range. Outline a spectrophotometric method that would accurately give the required data. Be specific about the solutions that you would use.

6. Using your equilibrium constant, check the assumption that all of the SCN^- was converted to $FeSCN^{2+}$ in tube 1. Assume that the value of $[FeSCN^{2+}]$ at equilibrium is an unknown quantity x. Express the values of $[Fe^{3+}]$ and $[SCN^-]$ at equilibrium based on the initial concentrations minus the amount that reacted to form $[FeSCN^{2+}]$. Solve the quadratic equation for x. What fraction of the thiocyanate ion is actually converted into the complex ion?

EXPERIMENT
19 Chemical Kinetics

APPARATUS

Clock with second hand, 25 × 200 mm test tubes, 600 mL beaker, thermometer.

REAGENTS

Soluble starch ("Vitex"), 0.010 M KI, 0.0010 M NaS$_2$O$_3$, 0.040 M KBrO$_3$, 0.10 M HCl, ice, 0.5 M (NH$_4$)$_2$MoO$_4$

INTRODUCTION

The rate law of the general reaction given by equation (1) is formulated by the general equation (2) where the square brackets refer to the molar concentrations of A and B, respectively.

$$a A + b B \longrightarrow c C + d D \tag{1}$$

$$\text{rate} = k\,[A]^m\,[B]^n \tag{2}$$

The exponents m and n are not necessarily equal to the coefficients a and b, respectively. The exponent m is the order of the reaction in terms of the concentration of A; the exponent n is the order of the reaction in terms of the concentration of B. The sum of the exponents $m + n$ is the order of the reaction as a whole. The constant k is known as the specific rate constant. The magnitude of k is a characteristic of the reaction and is temperature dependent.

The only way to determine quantitatively the influence of the concentration of any reacting species on the rate of reaction, and thus to determine the order of the reaction in terms of the concentration of each of the reactants, is by experiment. Calculation of the rate of the resulting reaction can be based on the rate of disappearance of one of the reactants or on the rate of appearance of one of the products.

This experiment studies the characteristics of the following reaction between iodide ion and bromate ion in acidic solution.

$$6\,I^- + BrO_3^- + 6\,H^+ \longrightarrow 3\,I_2 + Br^- + 3\,H_2O$$

The reaction occurs at a rate that can be conveniently studied several times within a laboratory period. The exponents of the following rate law can be determined by varying the concentrations of the three reactants.

$$\text{rate} = k\,[I^-]^m\,[BrO_3^-]^n\,[H^+]^p$$

The method used to study the rate of this reaction depends on the I$_2$ formed. A technique known as a clock reaction is used. The iodine formed reacts very rapidly with thiosulfate ion which is added initially to the reaction mixture. As a result, the concentration of I_2 is always near zero. However, the amount of thiosulfate is such that it eventually is exhausted. At that point the I_2 subsequently formed can be detected by its rapid reaction with a starch indicator that also has been added to the initial reaction mixture. The iodine forms an intensely blue colored complex with the starch. As a consequence we know how much time it takes for the reactants, regardless of their initial concentrations, to react to give a specific amount of I_2.

In this experiment, the concentrations of all of the reactants are varied, but a constant amount of thiosulfate is used. The time required for the iodine "clock" to signal the formation of a specific amount of I_2 is measured. That time is inversely proportional to the rate of the reaction.

$$\text{rate} \propto \frac{1}{\text{time}}$$

In order to avoid working with small numbers, the reciprocal of the time is multiplied by 1000. The relationship between the rates of the reactions of the various trials is unaltered. A series of reactions are studied in which the concentration of only one reactant at a time is changed with respect to a reference reaction. In this manner you can determine how each reactant affects the rate of the reaction. If the concentration of a reactant is double that in the reference reaction and the resultant rate is doubled, then the order of that reactant is the first power.

In addition to the effect of the concentration of the reactants on the rate of the reaction, you will determine the effect of temperature on the rate as well. Finally, the effect of a catalyst on the rate of reaction is determined.

PROCEDURE

Part A—Effect of Concentration on Reaction Rate

Obtain approximately 100 mL of 0.010 M KI, 0.0010 M $Na_2S_2O_3$, 0.040 M $KBrO_3$, and 0.10 M HCl in clean dry flasks or beakers. The volumes of the reagents required to prepare each of several trials are listed in Table 19-1. These amounts may be obtained by using either a 10.0 mL pipet or a 10 mL graduated cylinder.

Table 19-1

Reaction	Test tube 1			Test tube 2	
	0.0010 M $Na_2S_2O_3$	0.010 M KI	H_2O	0.040 M $KBrO_3$	0.10 M HCl
1	10 mL	10 mL	10 mL	10 mL	10 mL
2	10 mL	20 mL	0 mL	10 mL	10 mL
3	10 mL	10 mL	0 mL	20 mL	10 mL
4	10 mL	10 mL	0 mL	10 mL	20 mL

In order to start the reaction consistently and to accurately measure the rate of the reaction by the appearance of the iodine-starch complex, it is necessary to prepare combinations of reactants and the thiosulfate solution so that they can be mixed within a few seconds. Note that some of the reagents are mixed in a 25 × 200 mm test tube 1 and the others in a 25 × 200 mm test tube 2.

Place the KI and the thiosulfate (and water if indicated) in test tube 1 and the KBrO$_3$ and HCl in test tube 2. Then add a half microspatulaful of a soluble starch indicator to test tube 2. Determine the temperature of the contents of the test tubes. Ideally the temperature should be close to 20°C. The test tubes can be warmed or cooled by partially immersing them in a 600 mL beaker of water maintained at that temperature.

Pour the contents of test tube 2 into test tube 1 and swirl the contents to mix the reactants. Immediately note the time to the nearest second. Slowly agitate the contents of the test tube and be prepared to note the appearance of the blue iodine-starch complex. The time should be at least 1 minute and probably less than 3 minutes. Record the time on the Report Sheet.

Empty the contents of the test tube and rinse with distilled water. Repeat the experiment using the next set of concentrations of reactants. Note that the pipet or graduated cylinder used to deliver the various solutions should also be carefully rinsed. Remember to add the soluble starch to test tube 2 and each time a new reaction is run.

Part B—Effect of Temperature on Reaction Rate

Repeat the experiment two more times using the concentrations given for reaction mixture 1 in Part A at both 40°C and 0°C. Immerse the test tube in a 400 mL (or larger) beaker containing hot water or ice water for a sufficient time to change the temperature of the solutions. Determine the temperature of each test tube and then mix the contents as in Part A. However, keep the reaction test tube in the water bath to maintain the temperature. Record the time required for the color of the iodine-starch complex to appear.

Part C—Effect of a Catalyst on Reaction Rate

Prepare an experiment using the concentrations given for reaction mixture 1 in Part A at 20°C. However, this time add one drop of 0.5 M (NH$_4$)$_2$MoO$_4$ (ammonium molybdate) to test tube 2 along with the starch. Add the contents of test tube 2 to test tube 1 and agitate to rapidly mix the contents. Record the time required for a color change.

DATA ANALYSIS

Concentrations of solutions

In each reaction tube, the concentration of each reagent is not equal to original concentration of the stock solutions. They each are less because mixing with another solution results in a dilution. Note that mixing solutions of each test tube also dilutes all reagents. The concentration after mixing can be calculated from the dilution formula.

$$M_1 V_1 = M_2 V_2 \qquad\qquad (6)$$

M_1 and V_1 refer to the molarity and the volume of a species in the stock solution, whereas M_2 and V_2 refer to the molarity and volume after mixing both solutions. V_2 is the final total volume of the reaction mixture.

Order of the Reaction

The order of the reaction with respect to each reactant is determined by comparing the relative rates of two reactions under conditions where the concentrations of all reactants except one are identical. Because there is only one difference, the ratio of the rates of the reactions must be equal to a power of the ratio of the differing concentrations. Assume that the concentrations of reactants

A, B, and C in reaction 1 are x_1, y_1, and z_1, respectively and are x_2, y_1, and z_1, respectively for reaction 2. Write the equations for the general rate law.

$$\text{rate 1} = k\,(x_1)^m\,(y_1)^n\,(z_1)^p$$

$$\text{rate 2} = k\,(x_2)^m\,(y_1)^n\,(z_1)^p$$

The ratio of the two rates is equal to the ratio of the quantities x_1 and x_2 raised to the *m* power because the other concentration terms cancel.

$$\frac{\text{rate 1}}{\text{rate 2}} = \left[\frac{x_1}{x_2}\right]^m$$

If the ratio of the rates are 1:2 and the ratio of the concentrations is 1:2, then the value of m is 1. For the same ratio of concentrations, a reaction where m = 2 would have a 1:4 ratio of rates. If the ratio of the rates is 1:1, within experimental error, then m = 0.

Applying the same method to each of the possible reactants, you can determine the values of the exponents *n* and *p*. Note that for each determination you must choose a different set of two experiments so that the appropriate concentration, of the other two reactants are equal.

Effect of Temperature

The reactions at the three temperatures selected involve the same concentrations of reactants. For temperatures T_1 and T_2, expressed in kelvins, the rate expressions are

$$\text{rate}\,(T_1) = k(T_1)\,(x_1)^m\,(y_1)^n\,(z_1)^p$$

$$\text{rate}\,(T_2) = k(T_2)\,(x_1)^m\,(y_1)^n\,(z_1)^p$$

where $k(T_1)$ and $k(T_2)$ are the specific rate constants at the corresponding temperatures. Because the initial concentrations are the same, the ratio of the rates is equal to the ratio of the specific rate constants. Determine the ratio of the data at 40°C to 20°C and for the data at 20°C to 0°C.

$$\frac{\text{rate}\,(T_1)}{\text{rate}\,(T_2)} = \frac{k(T_1)}{k(T_2)}$$

Effect of a Catalyst on Reaction Rate

The reactions for the normal reaction and the catalyzed reaction are done at the same temperature and using the same concentrations of reactants. The rate expressions are

$$\text{rate (catalyzed)} = k(\text{catalyzed})\,(x_1)^m\,(y_1)^n\,(z_1)^p$$

$$\text{rate (uncatalyzed)} = k(\text{uncatalyzed})\,(x_1)^m\,(y_1)^n\,(z_1)^p$$

The ratio of the rates is equal to the ratio of the specific rate constants for the catalyzed and uncatalyzed reactions.

$$\frac{\text{rate (catalyzed)}}{\text{rate (uncatalyzed)}} = \frac{k(\text{catalyzed})}{k(\text{uncatalyzed})}$$

Report Sheet

Name _____ Instructor/Section _____ Date _____

Part A—Order of the Reaction

Reaction Mixture	Time t (sec)	Relative Rate 1000/t	Concentrations in Reaction Mixture		
			I^-	BrO_3^-	H^+
1	_____	_____	_____	_____	_____
2	_____	_____	_____	_____	_____
3	_____	_____	_____	_____	_____
4	_____	_____	_____	_____	_____

Order of Reactants I^- _____ BrO_3^- _____ H^+ _____

Part B—Effect of Temperature

Temperature (kelvins)	Time t (sec)	Relative Rate 1000/t	Concentrations in Reaction Mixture		
			I^-	BrO_3^-	H^+
20°C	_____	_____	_____	_____	_____
40°C	_____	_____	_____	_____	_____
0°C	_____	_____	_____	_____	_____

Ratio of rate constants k(40°C)/k(20°C)   _____

Ratio of rate constants k(20°C)/k(0°C)   _____

Part C—Effect of Catalyst

Reaction	Time t (sec)	Relative Rate 1000/t	Concentrations in Reaction Mixture		
			I^-	BrO_3^-	H^+
uncatalyzed	_____	_____	_____	_____	_____
catalyzed	_____	_____	_____	_____	_____

Attach sheets showing your calculations and answer the questions on the following sheet.

QUESTIONS

1. How much time would it take for the reaction in Part A to occur if the concentration of I^- were tripled compared to mixture 1. If the concentration of BrO_3^- were tripled? If the concentration of H^+ were tripled?

2. Does the amount of starch added to test tube 2 affect the determination of the kinetic order of the reaction?

3. At what time would the blue starch-iodine complex appear in reaction 1 if the concentration of sodium thiosulfate were 0.0015 M?

4. At what time would the blue starch-iodine complex appear in reaction 1 if the concentration of potassium iodide solution used were 0.020 M?

EXPERIMENT

20   Acids and Bases

APPARATUS

25 mL buret, ringstand, buret clamp, 125 mL Erlenmeyer flasks, 10 mL volumetric pipet, test tubes, stirring rods, spatula.

REAGENTS

0.1 M HCl, 0.1 M NaOH, 0.1 M CH_3CO_2H, 0.1 M NaCl, saturated Na_2CO_3, saturated $NaHCO_3$, 2 M aqueous NH_3, standardized NaOH (close to 0.1000 M), phenolphthalein indicator solution, fruit punch, cranberry juice, solid $NaHCO_3$, litmus paper, white vinegar, standardized unknown solutions of acetic acid.

INTRODUCTION

According to the Brønsted-Lowry definition, acids are proton donors and bases are proton acceptors. A base then must have a pair of electrons that can form a bond with a proton. Consider the reaction of HCl with water.

$$HCl + H_2O \longrightarrow H_3O^+(aq) + Cl^-(aq)$$

The HCl, which acts as an acid, is a proton donor and the H_2O, which acts as a base, is a proton acceptor. Because HCl is an excellent donor of protons, the reaction essentially goes to completion and HCl is classified as a strong acid.

A weak acid, such as acetic acid, is a   poor donor of protons. Acetic acid donates protons to water only to a limited extent. The concentration of the undissociated acid is much greater than that of acetate ion.

$$CH_3CO_2H(aq) + H_2O \rightleftharpoons CH_3CO_2^-(aq) + H_3O^+(aq)$$

Hydroxide salts of Group IA metals are strong bases. These salts dissociate completely in water to give hydroxide ions.

$$NaOH + H_2O \longrightarrow Na^+(aq) + OH^-(aq)$$

Weak bases are substances that yield only small concentrations of hydroxide ion. Ammonia is a common weak base. The nitrogen atom has an electron pair that accepts a proton from water.

$$H_2O + NH_3(aq) \rightleftharpoons OH^-(aq) + NH_4^+(aq)$$

However, in an aqueous solution of ammonia, the predominant species is ammonia. The hydroxide and ammonium ions are present in low concentrations.

Hydrolysis of Salts

Acidic and basic solutions may be produced from the reactions of salts with water. Such a reaction is called hydrolysis. Salts consisting of cations of Group IA and IIA metals and anions derived from strong acids produce neutral solutions when they are dissolved in water.

$$KCl + H_2O \longrightarrow K^+(aq) + Cl^-(aq)$$

However, aqueous solutions of salts of Group IA and IIA cations and anions derived from weak acids are basic. The anion is partially hydrolyzed in a reaction where the anion acts as a base and accepts a proton from water. Acetate salts behave in this manner.

$$CH_3CO_2^-(aq) + H_2O \rightleftharpoons CH_3CO_2H(aq) + OH^-(aq)$$

Certain salts partially hydrolyze to give acidic solutions. These salts have cations that can donate a proton to water. Ammonium salts behave in this manner.

$$NH_4^+(aq) + H_2O \rightleftharpoons NH_3(aq) + H_3O^+(aq)$$

Indicators

An indicator is an organic dye that can be used to test a solution to determine if it is acidic or basic. The indicator has one color in an acidic solution and a different color in a basic solution. An indicator is added to a sample of an unknown solution and the color is noted. If the color is that of the acid form of the indicator, the unknown solution is acidic. Similarly a basic solution would convert the indicator to its basic form and a different color would result.

Litmus is a complex mixture of organic compounds. This substance usually is impregnated into absorbent paper. Litmus paper comes in two colors: blue and red. Blue litmus paper turns red on contact with an acidic solution (pH <7). Red litmus paper turns blue on contact with a basic solution (pH >7). Thus, for litmus, red is the acidic color and blue is the basic color.

Phenolphthalein is another dye used as an indicator. Phenolphthalein is dispensed as a solution in ethanol. It is colorless in acidic solutions and bright pink in basic solutions. Before titrating an acidic solution, it is treated with a few drops of phenolphthalein indicator, which remains colorless. During the titration, a basic solution is added to the original mixture. The mixture remains colorless (acidic) as long as the total amount of base added is not sufficient to neutralize all of the acid. When the exact amount of base required to neutralize the acid has been added, the next drop of basic solution turns the original mixture from acidic to basic and the phenolphthalein turns pink (basic color). This color change is the signal to stop adding base. The amount of base added equals the amount of acid originally present in the mixture.

Natural Indicators

Some natural products have pigments that are acid-base indicators. The color of some flowers is due to anthocyanins, which are red in acids, violet when neutral, and blue in basic solutions. Thus, anthocyanin is responsible for the color of both blue cornflowers and red roses; only the pH is different.

Color is important to the food industry where consumer acceptance of a product often depends on its appearance. Food processing may result in undesirable color changes . For this reason artificial colorings are added to many commercial food products.

Vinegar

Commercial vinegar is prepared by fermenting apple juice. Various pigments in the apples give natural vinegar its brown color. Acetic acid, which is colorless, is the ingredient responsible for vinegar's sour taste. Colorless vinegar can be prepared by distilling the acetic acid from natural vinegar, leaving the brown ingredients behind. In this experiment you will analyze a sample of colorless vinegar that has been prepared by mixing pure acetic acid with water. The amount of acetic acid in your sample will be determined by titrating it with a standardized solution of sodium hydroxide.

Titration

The exact concentration of an acid or base in solution is determined by titration. In an acid-base titration, a known volume of the acid whose concentration is to be determined is placed in a flask. Base is then added to the acid sample until the number of moles of OH^- is exactly equal to the number of moles of H^+ of the acid solution. The concentration of the H^+ in the acid sample can then be determined from the equation

$$M_{acid} \ V_{acid} \ = \ M_{base} \ V_{base}$$

where M_{acid} and V_{acid} refer to the molarity of the H^+ and the volume of the acid solution, respectively, and M_{base} and V_{base} refer to the molarity of the OH^- and the volume of the base solution, respectively.

The exact point of neutralization of the acid by the base is determined by the use of an indicator that is added to the reaction mixture. In this experiment, you will titrate colorless vinegar with an NaOH solution of known concentration. At the equivalence point the reaction mixture will consist of sodium acetate and water, a slightly basic solution . For this reason you will use phenolphthalein, which changes from colorless in acid solution to pink at a pH of 9.

PROCEDURE

Part A—Litmus Paper

Blue litmus paper turns red upon contact with an acid solution (pH < 7). Red litmus paper turns blue on contact with a basic solution (pH > 7). Solutions that are nearly neutral do not change the color of litmus paper. Test the following solutions with each color of litmus paper and characterize each solution as acidic, basic, or neutral. Record your results on the Report Sheet.

0.1 M HCl (hydrochloric acid)
0.1 M NaOH (sodium hydroxide)
0.1 M CH_3CO_2H (acetic acid)
0.1 M NaCl (sodium chloride)
saturated Na_2CO_3 (sodium carbonate)
2 M NH_3 (ammonia)

Part B—Titration of Artificial Vinegar

The quality of your titration results will depend upon careful technique. Review in detail Sections I, "Reading a Meniscus" (pages 17-18), and J, "Volumetric Analysis" (pages 18-22).

In a stoppered flask obtain about 75 mL of standardized 0.1 M sodium hydroxide solution. Record the exact molarity (read from the bottle) on your Report Sheet to four significant figures. Clean a 25 mL buret with warm soapy water and rinse it thoroughly with distilled water. The buret is properly cleaned when a solution flows freely on the inside walls and does not form droplets. After thoroughly cleaning the buret, rinse it with about 5 mL of the standard sodium hydroxide solution. Mount the buret in a clamp on a ring stand as shown in Figure 24 on page 20. Now carefully fill the buret with standard sodium hydroxide solution, and drain the buret to just below the 0 mL mark. Record the reading on your Report Sheet.

Obtain a vinegar sample from your instructor. Pipet 10.00 mL of your vinegar sample into an Erlenmeyer flask. Add 2 drops of phenolphthalein indicator to the vinegar. Place the Erlenmeyer flask under the tip of the buret, and add the sodium hydroxide while swirling the flask. As the NaOH is added, a transient pink color that disappears on swirling is observed.

The color disappearance slows down as the titration proceeds, and consequently the addition of sodium hydroxide should be made dropwise near the endpoint. The titration is complete when the first permanent pink tinge appears. Record the final reading of the buret on the Report Sheet. Repeat the titration in order to obtain a duplicate determination. Perform a third titration if the difference in the volumes of base used in the first two trials is greater than 0.2 mL. Calculate the M_{acid} in the vinegar.

Part C—Natural Indicators

Using litmus paper, test the acidity of each of the following solutions. Record your data on the Report Sheet.

1. 3 mL of fruit punch
2. 3 mL of cranberry juice
3. 3 mL of a saturated solution of sodium bicarbonate ($NaHCO_3$)

Take approximately 3 mL of fruit punch and a separate 3 mL sample of cranberry juice in two medium test tubes. Obtain a few grams of solid sodium bicarbonate on a clean piece of filter paper and, using your spatula, slowly add small amounts of $NaHCO_3$ to each sample. If a color change is observed, immediately discontinue adding the $NaHCO_3$. (If sufficient bicarbonate is added to produce a saturated solution, it is safe to assume that no change in color will occur.) Record your observations on the Report Sheet. If the addition of $NaHCO_3$ caused the color of the solution to change, add a small amount of colorless vinegar until the solution is slightly acidic. Record your observations on the Report Sheet.

Report Sheet

| Name | Instructor/Section | Date |

Part A—Litmus Paper

Solution	Effect on Litmus	Conclusions
0.10 M HCl		
0.10 M NaOH		
0.10 M CH$_3$CO$_2$H		
0.10 M NaCl		
Saturated Na$_2$CO$_3$		
2 M NH$_3$		

Explain by equations why a solution of sodium carbonate changes the color of litmus paper.

Part B—Titration

	Trial 1	Trial 2	Trial 3
Molarity of standard NaOH			
Volume of vinegar sample (mL)			
Initial buret reading (mL)			
Final buret reading (mL)			
Volume of NaOH required for neutralization (mL)			
Molarity of acetic acid in vinegar			
Average molarity of acetic acid			

Part C—Natural Indicators

Solution	Effect on Litmus
Fruit punch	_____
Cranberry juice	_____
Saturated $NaHCO_3$	_____

Fruit punch + $NaHCO_3$

    Is a gas produced? _____

    Is there a color change? _____ If so, from to _____ to _____

                                    If so, effect of adding vinegar _____

Cranberry juice + $NaHCO_3$

    Is a gas produced? _____

    Is there a color change? _____ If so, from to _____ to _____

                                    If so, effect of adding vinegar _____

Attach sheets with sample calculations and the answers to the questions sheet on the following pages.

QUESTIONS

1. Write the expected products in each of the following reactions.

    a. $HBr + LiOH \longrightarrow$

    b. $HI + Ba(OH)_2 \longrightarrow$

    c. $H_2S + H_2O \longrightarrow$

    d. $NH_3 + H_2O \longrightarrow$

2. Examine a table of K_a values and list three weak acids.

3. Would you expect RbOH to be a strong base or a weak base?

4. Why is ammonia considered a weak base?

5. Predict whether solutions formed from dissolving the following salts in water will be acidic, basic, or nearly neutral.

    a. LiCN

    b. $KClO_4$

    c. NH_4Br

    d. CH_3CO_2Na

    e. $NaNO_3$

6. Would an indicator that changes at pH 7 be appropriate for the titration of vinegar with KOH?

7. A sample of ammonia is titrated with HCl. Will the pH of the solution at the end point be 7? Explain why or why not.

8. Could a 5% solution of HCl be used as a substitute for vinegar in a salad dressing? Why or why not?

9. How do you explain the differences in the effects of adding $NaHCO_3$ to cranberry juice and fruit punch?

10. Assuming the vinegar solution has a density of 1.01 g/mL, calculate the percent by weight of acetic acid (CH_3CO_2H) in your sample of vinegar. How does that value compare with 5%, the value for the % acetic acid in normal vinegar?

EXPERIMENT

21   Redox Reactions of the Halogens

APPARATUS

10 × 75 mm micro test tubes, corks for test tubes, rack for test tubes, medicine dropper, beakers.

REAGENTS

Solids. sodium chloride, sodium bromide, sodium iodide, sodium bromate.
Solutions. chlorine water, bromine water, iodine water, 9 M sulfuric acid (H_2SO_4), small dropper bottles containing 0.1 M solutions of sodium chloride (NaCl), sodium bromide (NaBr), sodium iodide (NaI), silver nitrate ($AgNO_3$), and 6 M nitric acid (HNO_3), 1,2-dichloroethane (CH_2ClCH_2Cl).

INTRODUCTION

The halogens fluorine, chlorine, bromine, and iodine are the most distinctly nonmetallic family of the periodic chart. These elements have similar chemical properties that change according to their position in the periodic chart.

One of the common properties of the halogens is their oxidizing ability. This property is demonstrated by their reaction with metals. For example, chlorine reacts with metallic sodium.

$$2\ Na\ +\ Cl_2\ \longrightarrow\ 2\ NaCl$$

In this reaction, each chlorine atom gains one electron to give a chloride ion in the compound sodium chloride. The chlorine atom has been reduced from its initial oxidation number of zero to an oxidation number of –1 in sodium chloride. Since chlorine is reduced, it is the oxidizing agent. You will have the opportunity to test the relative oxidizing abilities of chlorine, bromine, and iodine in this experiment.

The reduced forms of the halogens are halide ions that have the oxidation number –1. In contrast to the free halogens, the halide ions fluoride, chloride, bromide, and iodide are reducing agents.

Redox Reactions of the Halogens

The reducing ability of iodide ion can be demonstrated by the reduction of copper(II) ions to copper(I) in aqueous solutions.

$$2\,I^- \ + \ 2\,Cu^{2+} \ \longrightarrow \ 2\,Cu^+ \ + \ I_2$$

During the reaction, each iodide ion loses one electron to give an iodine atom in the I_2 molecule. The iodide ion has been oxidized from its initial oxidation number of –1 to an oxidation number of 0 in elemental iodine. In this example, the iodide ion loses an electron; iodide ion is oxidized and acts as a reducing agent. You will compare the relative reducing properties of the chloride, bromide, and iodide ions.

In the course of this experiment you will need to identify the free halogens. Bromine and iodine vapors are red-brown and violet, respectively, and are readily visible. Chlorine gas is yellow-green, but when present in small amounts it is often difficult to see. Chlorine does have a characteristic strong odor. The odor of liquid household bleach is due to chlorine gas.

When silver nitrate is added to a solution containing halide ions, a precipitate of the silver halide is formed. These halide compounds have characteristic colors. Silver chloride is pure white, silver bromide is off-white, and silver iodide is yellowish.

With the exception of fluorine, the halogen elements form compounds with the oxidation numbers +1, +3, and +5. In Part D you will study a compound of bromine with the oxidation number +5, sodium bromate ($NaBrO_3$). Finally, an unknown will be identified. Your unknown will be one of the following salts: NaCl, NaBr, NaI, or $NaBrO_3$.

PROCEDURE

CAUTION: Halogen elements cause skin burns. Chlorine and bromine vapors are caustic. Do not inhale them.

WARNING: 1,2-Dichloroethane *hexane* **is flammable. It must be kept away from all open flames such as Meker burners. Discarded 1,2-dichloroethane should be placed in the special waste can for organic solvents.**

CAUTION: Both nitric acid and sulfuric acid are a strong corrosive acids that will cause severe skin burns. Immediately wash any affected areas with large quantities of water. Notify your instructor and seek a medical evaluation of the chemical burn.

Part A—Properties of Halogens in Solution   *hood*

hexanes The halogens dissolve in solvents such as water, chloroform ($CHCl_3$), and 1,2-dichloroethane ($CH_2Cl—CH_2Cl$). Each solution has a characteristic color. In this experiment, 1,2-dichloroethane is added to a given reaction mixture to intensify the color of the halogen in solution. This intensification results from the greater solubility of halogens in 1,2-dichloroethane than in water. 1,2-Dichloroethane is not soluble in water. A mixture containing 1,2-dichloroethane and water forms two layers: water on the top and 1,2-dichloroethane on the bottom. If a halogen is present in this mixture, the halogen will dissolve almost totally in the bottom 1,2-dichloroethane layer. The color of this bottom layer can be used to identify the halogen.

162

Place 10 drops each of chlorine water, bromine water, and iodine water into separate labeled micro test tubes. Note the color of each aqueous solution. Place 20 drops (1 mL) of 1,2-dichloroethane in each tube. Stopper and shake each tube. Place each tube in front of a plain white paper and record the color of each 1,2-dichloroethane layer on the Report Sheet. Save the tubes for later comparisons.

Part B—Oxidizing Abilities of the Halogens

Place a quantity about the size of a grain of rice of NaBr into each of two labeled micro test tubes. Place a similar quantity of NaI into each of two additional labeled micro test tubes. Dissolve each sample in about 20 drops of water. Now add 10 drops of chlorine water to one tube containing the NaBr solution and to one tube containing the NaI solution. Add 10 drops of bromine water to both remaining tubes. Add 20 drops of 1,2-dichloroethane to each of the four tubes, stopper each tube, and shake vigorously. Record on the Report Sheet the color of the 1,2-dichloroethane layer in each tube. Compare the tubes with the tubes saved from Part A and identify each product on the basis of color. If no reaction occurs, write N.R.

Part C—Silver Halides

To three labeled micro test tubes, add 7 drops of water and 3 drops of 6 M nitric acid (HNO_3). Add 5 drops of 0.1 M sodium chloride (NaCl) to the first tube, 5 drops of 0.1 M sodium bromide (NaBr) to the second, and 5 drops of 0.1 M sodium iodide (NaI) to the third. Then add 10 drops of 0.1 M silver nitrate ($AgNO_3$) to all three tubes. Record the shade of each silver halide precipitate. Save the tubes containing the precipitates for later comparisons.

Part D—Sodium Bromate: Bromine with the Oxidation Number +5

Clean and label three micro test tubes I, II, and III. Place a quantity of sodium bromate ($NaBrO_3$) about the size of a grain of rice in each tube and add 20 drops of water and 10 drops of 1,2-dichloroethane. To tubes I and II add 10 drops of 0.1 M sodium bromide solution (NaBr). Then to tubes I and III add 2 drops of 9 M sulfuric acid (H_2SO_4). Stopper each tube and shake vigorously. Record your observations on the Report Sheet. In which tube did the dichloroethane layer become colored? Identify the product of the reaction on the basis of this color. If no reaction occurs, write N.R.

Part E—Identification of an Unknown Containing NaI, NaBr, NaCl, or $NaBrO_3$

An unknown containing either NaI, NaBr, NaCl, or $NaBrO_3$ will be identified. The basis for the identification will be the chemical reactions studied above. To identify the unknown, a series of procedures from Parts A, B, C, and D should be followed. In each procedure the known substance is replaced by the unknown to be tested. For example, in procedure Part C silver nitrate is added to a solution of the unknown and the result is compared with those obtained previously by adding silver nitrate to each known. If performed properly, the result of this test on the unknown will be the same as that obtained for one of the knowns. The unknown thus is identified.

Obtain an unknown sample from the instructor and test it first for the bromate ion with the procedure in Part D. If the compound is not $NaBrO_3$, then test the unknown further by the procedures in Parts B and C. Remember to replace the known halogen-containing compound with your unknown when performing each test.

Report Sheet

Name	Instructor/Section	Date

Part A—Color of Halogens in Solution

Halogen	Water solution	1,2-Dichloroethane solution
Chlorine	_____	_____
Bromine	_____	_____
Iodine	_____	_____

Part B—Oxidizing Abilities of the Halides

Reactants	Color in 1,2-dichloroethane	Products of reaction
$NaBr + Cl_2$	_____	_____
$NaBr + Br_2$	_____	_____
$NaI + Cl_2$	_____	_____
$NaI + Br_2$	_____	_____

Part C—Color of Silver Halides

Compound	Color of Solid
AgCl	_____
AgBr	_____
AgI	_____

Part D—Sodium Bromate

Reactants	Observation	Products of reaction
$NaBrO_3$ + NaBr + H_2SO_4	_____	_____
$NaBrO_3$ + NaBr	_____	_____
$NaBrO_3$ + H_2SO_4	_____	_____

Part E—Identification of Unknown Salt

	Test performed	Observation	Conclusion
1	_____	_____	_____
2	_____	_____	_____
3	_____	_____	_____
4	_____	_____	_____
5	_____	_____	_____

Identity of unknown halogen salt _____

Answer the questions on the following page.

QUESTIONS

1. For each reaction that occurred in Part B identify the chemical that is oxidized and the chemical that is reduced. Also identify the chemical species that is acting as the oxidizing agent and the species that is the reducing agent.

2. Which of the following species is the weakest oxidizing agent: Cl_2, Br_2, I_2? Explain your reasoning.

3. Which of the following species is the strongest reducing agent: Cl^-, Br^-, I^- ? Explain your reasoning.

4. Explain why a reaction occurred in only one of the test tubes in Part D.

5. Calculate the oxidation numbers for bromine in $KBrO_3$ and in KBr. Show your work.

6. For the reaction between sodium bromide, sodium bromate, and sulfuric acid, identify the chemical that is oxidized and the one that is reduced. (The sulfate anion does not participate in any oxidation or reduction—it is just a spectator ion.)

7. Predict what would happen if sodium iodide, sodium iodate ($NaIO_3$), and sulfuric acid solutions were mixed with a small amount of 1,2-dichloroethane.

8. Fluorine does not form a fluorate ion, FO_3^-. Explain why.

22   Redox Reactions of Metals

APPARATUS

Test tubes 10 × 75 mm and 16 × 150 mm, test tube racks, 150 mL beaker, medicine dropper, 100 mL graduated cylinder, sandpaper.

REAGENTS

5 mm × 50 mm strips of Fe, Cu, Zn; 6 M HCl; 6 M NaOH; methanol; 0.1 M solutions of $CuSO_4$, $ZnSO_4$, $Fe(NH_4)_2(SO_4)_2$, KI, Na_2SO_3; 1% $KMnO_4$ solution; 10% $Na_2Cr_2O_7$ solution, 0.5 M $SnCl_2$ in 3 M HCl.

INTRODUCTION

In an oxidation-reduction reaction, one or more electrons are transferred from one substance to another. A simple example is the reaction between zinc metal and tin(II) ions:

$$Zn(s) + Sn^{2+}(aq) \longrightarrow Zn^{2+}(aq) + Sn(s)$$

The reaction between zinc metal and tin(II) ions is an example of a displacement reaction. The zinc metal displaces the tin ions from the solution. In this reaction the zinc metal loses two electrons. The zinc is oxidized since any substance that loses electrons is oxidized.

$$Zn(s) \longrightarrow Zn^{2+}(aq) + 2\ e^- \qquad \text{(oxidation)}$$

An oxidation reaction is accompanied by a reduction reaction (or gain of electrons). Here the electrons are taken up by the tin(II) ions to produce tin metal. The tin(II) ions are reduced.

$$Sn^{2+}(aq) + 2\ e^- \longrightarrow Sn(s) \qquad \text{(reduction)}$$

Adding the oxidation half-reaction to the reduction half-reaction yields the overall reaction.

$$Zn(s) \longrightarrow Zn^{2+}(aq) + 2\ e^- \qquad \text{(oxidation)}$$
$$Sn^{2+}(aq) + 2\ e^- \longrightarrow Sn(s) \qquad \text{(reduction)}$$

$$Sn^{2+}(aq) + Zn(s) \longrightarrow Zn^{2+}(aq) + Sn(s) \quad \text{(oxidation-reduction)}$$

Redox Reactions of Metals

When combining half-reactions, you must take care to balance the electrons. Consider combining the following two equations.

$$Zn(s) \longrightarrow Zn^{2+}(aq) + 2\,e^- \qquad\qquad \text{(oxidation)}$$

$$Ag^+(aq) + 1\,e^- \longrightarrow Ag(s) \qquad\qquad \text{(reduction)}$$

You must multiply everything in the silver (reduction) half-reaction by 2 so that two electrons are obtained for canceling the two electrons already in the zinc (oxidation) half-reaction.

$$Zn(s) \longrightarrow Zn^{2+}(aq) + 2\,e^- \qquad\qquad \text{(oxidation)}$$

$$2\,Ag^+(aq) + 2\,e^- \longrightarrow 2\,Ag(s) \qquad\qquad \text{(reduction)}$$

_____

$$2\,Ag^+(aq) + Zn(s) \longrightarrow Zn^{2+}(aq) + 2\,Ag(s)$$

Metals differ in their ability to displace other ions from solution. The electrochemical series is a listing of relative displacement ability for metals. Since zinc metal displaced tin(II) ions in the example given, one would not expect tin metal to displace zinc ions from solution. Thus the reaction will not proceed as written. Zinc will displace tin ions but tin will not displace zinc ions.

$$Zn^{2+}(aq) + Sn(s) \;\xrightarrow{\;\;\times\;\;}\; Sn^{2+}(aq) + Zn(s)$$

The evolution of hydrogen gas from the action of acid on a metal is a displacement reaction. Some metals can displace hydrogen; some cannot. Magnesium metal reacts with acid to release hydrogen gas.

$$Mg(s) + 2\,H^+(aq) \longrightarrow H_2(g) + Mg^{2+}(aq)$$

Lead metal cannot release hydrogen gas.

$$Pb(s) + 2\,H^+(aq) \;\xrightarrow{\;\;\times\;\;}\; H_2(g) + Pb^{2+}(aq)$$

You will perform several displacement reactions in this experiment.

Additional factors may complicate oxidation-reduction (redox) reactions. Redox reactions can be accompanied by gas evolution and precipitate formation. They may be affected by the presence of an acid or a base also. For example, iodide and iodate ions may be mixed without a reaction.

$$I^-(aq) + IO_3^-(aq) \longrightarrow \text{N.R.}$$

The addition of a third ingredient, acid, triggers a reaction. The colorless mixture of I^- and IO_3^- turns brown only after the addition of acid.

$$6\,H^+(aq) + 5\,I^-(aq) + IO_3^-(aq) \longrightarrow \underset{\text{(brown)}}{3\,I_2(aq)} + 3\,H_2O$$

Potassium permanganate ($KMnO_4$) is a violet crystalline solid and is violet in aqueous solution. When it is reduced, usually a precipitate of manganese dioxide (MnO_2) is formed.

MnO_2 is a dark brown sludge. A change from a violet solution to a dark brown precipitate offers an easy visual clue that the manganese has been reduced. The half-reactions for acid and base solutions are

$$3\ e^- + MnO_4^-(aq) + 4\ H^+(aq) \longrightarrow MnO_2(s) + 2\ H_2O \qquad \text{(in acid)}$$
$$\underset{\text{violet}}{} \qquad\qquad\qquad \underset{\text{dark brown}}{}$$

$$3\ e^- + MnO_4^-(aq) + 2\ H_2O\ (aq) \longrightarrow MnO_2(s) + 4\ OH^-(aq) \qquad \text{(in base)}$$
$$\underset{\text{violet}}{} \qquad\qquad\qquad\qquad \underset{\text{dark brown}}{}$$

Sodium dichromate ($Na_2Cr_2O_7$) is another compound that can be reduced easily; $Na_2Cr_2O_7$ is a bright orange crystalline solid that forms a yellow-orange solution. In the presence of acid the dichromate can be reduced to a green chromium(III) ion, $Cr^{3+}(aq)$.

$$6\ e^- + Cr_2O_7^{2-}(aq) + 14\ H^+(aq) \longrightarrow 2\ Cr^{3+}(aq) + 7\ H_2O \ \ \text{(in acid)}$$
$$\underset{\text{yellow-orange}}{} \qquad\qquad\qquad \underset{\text{green}}{}$$

In this experiment, $Na_2Cr_2O_7$ and $KMnO_4$ will be mixed with other substances that can be oxidized. Several possible oxidation half-reactions are

$$2\ I^-(aq) \longrightarrow I_2(aq) + 2\ e^-$$
$$2\ S_2O_3^{2-}(aq) \longrightarrow S_4O_6^{2-}(aq) + 2\ e^-$$
$$SO_3^{2-}(aq) + H_2O \longrightarrow SO_4^{2-}(aq) + 2\ H^+(aq) + 2\ e^-$$
$$H_2O + CH_3OH(aq) \longrightarrow HCO_2H(aq) + 4\ H^+(aq) + 4\ e^- \text{ (acidic solution)}$$
$$5\ OH^-(aq) + CH_3OH(aq) \longrightarrow HCO_2^-(aq) + 4\ H_2O + 4\ e^- \text{ (basic solution)}$$

All products in these reactions are colorless except $I_2(aq)$, which is yellow to dark brown depending on concentration.

PROCEDURE

Part A—Displacement Reactions

Obtain two strips of each of the following metals: Zn, Cu, Fe. Polish each strip with sandpaper and dip each into a 150 mL beaker containing 6 M HCl.

CAUTION: Avoid skin contact with HCl. Flush affected areas with a large amount of water and inform your instructor.

Note which metals react to form bubbles of hydrogen gas. Record your observations on the Report Sheet. Quickly remove the strips from the acid and rinse them with distilled water.

Obtain six clean 16 × 150 mm test tubes and add 25 mL of 0.1 M $CuSO_4$ to two of them, 25 mL of 0.1 M $ZnSO_4$ to two of them, and 25 mL of 0.1 M $Fe(NH_4)_2(SO_4)_2$ to the last two.

Redox Reactions of Metals

Test for displacement reactions between each of the three metal-metal ion pairs. Run six tests . Dip the metal strips into the 0.1 M solutions according to Table 22-1.

Table 22-1

Test	Metal strip	0.1 M solution
1	Cu	$Fe(NH_4)_2(SO_4)_2$
2	Fe	$CuSO_4$
3	Cu	$ZnSO_4$
4	Zn	$CuSO_4$
5	Fe	$ZnSO_4$
6	Zn	$Fe(NH_4)_2(SO_4)_2$

After 15 minutes, observe the metal strips and record your observations on the Report Sheet. On the basis of your observations, decide which metal in each pair of tests is the more reactive. Finally, take the strip of Zn metal from the $Fe(NH_4)_2(SO_4)_2$ solution and dip it into a solution containing 0.5 M $SnCl_2$ and 3 M HCl. Write the equations for two chemical reactions that occurred.

Part B—Oxidation-Reduction Involving Complex Ions

In separate 10 × 75 mm test tubes, add 5 drops of each of the following solutions: 0.1 M KI (tube 1), 0.1 M Na_2SO_3 (tube 2), methanol (tube 3), and methanol (tube 4). Set aside tube 4 for later use.

To tube 1 add 1% $KMnO_4$ dropwise. Stop adding when a permanent change has occurred. Do not exceed 10 drops. If no reaction has occurred, add 2 drops of 6 M HCl. Record your observations on the Report Sheet. Repeat for tubes 2 and 3.

To tube 4 containing 5 drops of methanol, add 10 drops of 1% $KMnO_4$; then add 2 drops of 6 M NaOH. Record your observations on the Report Sheet.

In separate 10 × 75 mm test tubes, add 5 drops each of 0.1 M KI and 0.1 M Na_2SO_3. Add 10% $Na_2Cr_2O_7$ dropwise to each tube until a permanent change has occurred. Do not exceed 10 drops. If no reaction occurs, add 2 drops of 6 M HCl. Record your observations on the Report Sheet.

Report Sheet

Name	**Instructor/Section**	**Date**

Part A—Displacement Reactions

Reactivity of metals with 6 M HCl

Metal	Observations
Zn	
Cu	
Fe	

Reactivity of metals with metal ion solutions

Test	Metal strip	0.1 M solution	Observation	Most reactive metal
1	Cu	$Fe(NH_4)_2(SO_4)_2$		
2	Fe	$CuSO_4$		
3	Cu	$ZnSO_4$		
4	Zn	$CuSO_4$		
5	Fe	$ZnSO_4$		
6	Zn	$Fe(NH_4)_2(SO_4)_2$		

Complete and balance the following equations. If no reaction occurred, then write N.R.

$Zn + HCl \longrightarrow$

$Cu + HCl \longrightarrow$

$Fe + HCl \longrightarrow$

$Cu + Fe^{2+} \longrightarrow$

$Fe + Cu^{2+} \longrightarrow$

$Cu + Zn^{2+} \longrightarrow$

$Zn + Cu^{2+} \longrightarrow$

$Fe + Zn^{2+} \longrightarrow$

$Zn + Fe^{2+} \longrightarrow$

$Zn + Sn^{2+} \longrightarrow$

$Zn + H^+ \longrightarrow$

Part B—Oxidation-Reduction Involving Complex Ions

Addition of 1% $KMnO_4$

| Test | Reactant | Observations | |
		No acid or base	With acid or base
1	0.1 M KI	_____	_____
2	0.1 M Na_2SO_3	_____	_____
3	CH_3OH	_____	_____
4	CH_3OH	_____	_____

Write balanced chemical equations.

1. $KMnO_4 + KI \longrightarrow$

2. $KMnO_4 + Na_2SO_3 \longrightarrow$

3. $KMnO_4 + CH_3OH \longrightarrow$

4. $KMnO_4 + CH_3OH \longrightarrow$

Addition of 10% $Na_2Cr_2O_7$

| Test | Reactant | Observations | |
		No acid	With acid
1	0.1 M KI	_____	_____
2	0.1 M Na_2SO_3	_____	_____

Write balanced chemical equations.

1. $Na_2Cr_2O_7 + KI \longrightarrow$

2. $Na_2Cr_2O_7 + Na_2SO_3 \longrightarrow$

Answer the questions on the following page.

QUESTIONS

1. Calculate the oxidation number of the indicated element in each of the following species.

    a. Cr in $Cr_2O_7^{2-}$

    b. Cr in CrO_4^{2-}

    c. Mn in K_2MnO_4

    d. Mn in MnO_4^{2-}

    e. P in H_3PO_3

    f. N in NO_2^-

2. Consider the following equation and split it into two half-reactions.

$$2\ S_2O_3^{2-}(aq)\ +\ I_2(aq)\ \longrightarrow\ S_4O_6^{2-}(aq)\ +\ 2\ I^-(aq)$$

3. Iron(III) ions and iodide ions react according to the following equation. Split the equation into half-reactions.

$$2\ Fe^{3+}(aq)\ +\ 2\ I^-(aq)\ \longrightarrow\ I_2(aq)\ +\ 2\ Fe^{3+}(aq)$$

4. Would you expect the following reaction to occur? (*Hint:* Consider the reactions in Questions 2 and 3.)

$$2 \ Fe^{3+}(aq) \ + \ S_2O_3^{2-}(aq) \longrightarrow 2 \ Fe^{3+}(aq) \ + \ S_4O_6^{2-}(aq)$$

EXPERIMENT

23 Qualitative Analysis of Cations and Anions

APPARATUS

Sixteen 10 × 75 mm micro test tubes, gummed labels, test tube rack for micro test tubes, medicine dropper, 10 mL graduated cylinder, cork stoppers for micro test tubes.

REAGENTS

0.1 M aqueous solutions of $CaCl_2$, $BaCl_2$, $Zn(NO_3)_2$, $AgNO_3$, Na_2SO_4, Na_2CrO_4, Na_2CO_3, NaI, NaCl, and Na_3PO_4; 6 M aqueous solutions of nitric acid (HNO_3) and hydrochloric acid (HCl).

INTRODUCTION

Qualitative analysis is based on determining the response of an ion to several reagents. No two ions behave exactly the same when exposed to a variety of reagents. For example, cations A^+ and B^+ may both give white precipitates with anion X^-, but only A^+ will give a white precipitate with Y^-. Thus, if both X^- and Y^- give white precipitates with an unknown, A^+ is present. If a precipitate occurs only with X^- but not Y^- then B^+ is present. This behavior is summarized in Table 23-1.

Table 23-1. Behavior of hypothetical cations A^+ and B^+ with test reagents X^- and Y^-

Ion	Test reagent	
	X^-	Y^-
A^+	White precipitate	White precipitate
B^+	White precipitate	Clear solution

In this experiment a limited number of cations and anions will be examined. Only a few test reagents will then be necessary. Observations will first be made on known materials. The data can then be used to identify an unknown. Generally, the procedure used should identify the cation or anion with the fewest possible reagents.

The experiment is divided into three parts. In Part A the cations Ca^{2+}, Ba^{2+}, Zn^{2+}, and Ag^+ will be studied with anions as test reagents. In part B the anions Cl^-, I^-, SO_4^{2-}, CO_3^{2-}, and PO_4^{3-} will be studied with cations as test reagents. In addition, HNO_3 will be used to distinguish further between the precipitates formed in the reaction of the anions and the test reagents.

In Part C of this experiment you will determine (a) the identity of a cation in one unknown solution and (b) the identity of an anion in another unknown solution.

PROCEDURE

Part A—Cation Analysis

Wash sixteen 10 × 75 mm test tubes and rinse them with distilled water. Place 1 mL (20 drops) of each of the cation solutions $CaCl_2$, $BaCl_2$, $Zn(NO_3)_2$, and $AgNO_3$ in labeled tubes. Add 1 mL (20 drops) of Na_2SO_4 as the anion reagent to each tube and gently mix the solutions by shaking the tube. Record your observations on your Report Sheet. Typical observations include a description of the type of precipitate and its color or in the case of a gas, its odor. Write the formula of any solid or gas formed.

Place another 1 mL of each of the four cation solutions in labeled tubes. Add 1 mL of Na_2CO_3 as the anion reagent to each tube. Record your observations and write the formula of each precipitate or gas formed.

Repeat this procedure with Na_2CrO_4 as the test reagent and then with NaI as the test reagent. Record your observations.

Part B—Anion Analysis

CAUTION: Nitric acid (HNO_3) should not be spilled or splashed out of the test tube. Promptly wash any acid spilled on your skin with large quantities of water. Immediately inform your instructor about your accident.

CAUTION: Avoid spilling silver nitrate ($AgNO_3$) on your skin. Although the skin discoloration (black) is not a health hazard, the result may be considered unsightly. When new skin is formed, the darkened skin will be discarded.

Wash ten 10 × 75 mm test tubes and rinse them with distilled water. Place 1 mL (20 drops) of each of the anion solutions NaCl, NaI, Na_2SO_4, Na_2CO_3, and Na_3PO_4 in five labeled test tubes. Add 1 mL (20 drops) of $AgNO_3$ solution to each tube, and shake the tube gently. Record your observations. Now add to each mixture a third component, 1 mL of dilute HNO_3, and again agitate the tube. Record your observations.

Place 1 mL of the five anion solutions in five labeled test tubes. Add 1 mL of $BaCl_2$ solution to each tube, and record your observations. Add 1 mL of 6 M HCl to each test tube, and record your observations.

Part C—Unknown Analysis

Obtain 5 mL of an assigned unknown cation solution and 5 mL of an assigned unknown anion solution. Using test reagents, determine the identity of the required ion in each solution. Record your observations. The known tests described in Parts A and B can be repeated to help resolve doubts about the identity of the unknown. Record your observations.

Report Sheet

Part A

Ion	Test Reagent			
	Na_2SO_4	Na_2CO_3	Na_2CrO_4	NaI
Ca^{2+}				
Ba^{2+}				
Zn^{2+}				
Ag^+				

Part B

Ion	Test Reagent			
	$AgNO_3$	Addition of HNO_3	$BaCl_2$	Addition of HCl
Cl^-				
I^-				
SO_4^{2-}				
CO_3^{2-}				
PO_4^{3-}				

Part C

Cation Unknown Number	Test Reagent			
	Na_2SO_4	Na_2CO_3	Na_2CrO_4	NaI

The unknown cation is _____

Anion Unknown Number	Test Reagent			
	$AgNO_3$	Addition of HNO_3	$BaCl_2$	Addition of HCl

The unknown anion is _____

Answer the questions on the following pages.

QUESTIONS

1. Complete and balance each of the following equations.

   a. $Zn(NO_3)_2 + Na_2CrO_4 \longrightarrow$

   b. $BaCl_2 + Na_2CO_3 \longrightarrow$

   c. $Ag_3PO_4 + HCl \longrightarrow$

   d. $BaCl_2 + Na_3PO_4 \longrightarrow$

   e. $CaCO_3 + HBr \longrightarrow$

   f. $AgClO_4 + NaBr \longrightarrow$

2. Which cation in this experiment gives the most unique results that would allow you to clearly identify it? Explain.

3. What common reaction occurs when acid is added to the tube containing $AgNO_3$ and Na_2CO_3 and the tube containing $BaCl_2$ and Na_2CO_3?

4. Explain the differences between the results of adding HNO_3 to solutions of $AgNO_3$ with CO_3^{2-} and PO_4^{3-}.

5. Predict the reaction, if any, that would occur if $AgNO_3$ and KBr were mixed. Would any reaction occur if HNO_3 was then added?

6. Predict the reaction, if any, that would occur if $Sr(NO_3)_2$ and Na_2CO_3 were mixed. Would any reaction occur if HBr was then added?

7. Predict the reaction, if any, that would occur if $Cd(NO_3)_2$ and K_2CrO_4 were mixed.

8. Silver acetate ($AgC_2H_3O_2$) is insoluble in water but is soluble in dilute nitric acid. Write equations explaining these observations.

EXPERIMENT

24   Molecular Models

MATERIALS

Molecular model set containing carbon with four tetrahedrally spaced holes, hydrogen with one hole, and bromine with one hole; two 1 cm strips of masking tape or two peelable labels (1 cm × 1 cm).

INTRODUCTION

In this experiment you will build molecular models and examine the structures of some hydrocarbon molecules. These models can aid you in drawing structural formulas and recognizing isomers. Two types of isomers, chain and positional, will be examined.

When building molecular models, it is possible to form two models that appear different but in fact represent the same compound. The same error may be made in drawing structural formulas. The three models shown in Figure 24-1 represent different conformations of the same isomer, pentane. Conformations result from rotation about single bonds. By convention, chemists choose to write structural formulas with the carbon chains in a straight line rather than as a bent chain. Thus, you should arrange your models in fully extended chains.

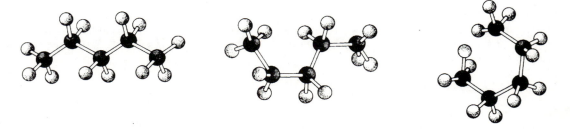

Figure 24-1. Conformations of pentane.

Molecular models also are useful in seeing the equivalence of carbon atoms within the molecule. The equivalence becomes evident from molecular formulas as well. In butane (Figure 24-2), the two terminal carbon atoms are identical in their chemical environment and are equivalent. The six bonded hydrogen atoms are also equivalent.

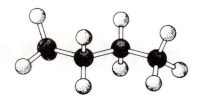

Figure 24.2 Molecular model of butane.

The two interior carbon atoms in butane are equivalent. The four hydrogen atoms bonded to the interior carbon atoms are equivalent also. These relationships can also be seen in structural formulas.

Equivalent carbon atoms

C—C—C—C

Equivalent carbon atoms

Equivalent hydrogen atoms

CH_3—CH_2—CH_2—CH_3

Equivalent hydrogen atoms

Recognition of the equivalence or lack of equivalence among carbon atoms and their bonded hydrogen atoms is important in writing positional isomers of related compounds. Consider the replacement of a single hydrogen atom in butane with a bromine atom. Only two possibilities exist. One is the replacement of a hydrogen atom bonded to either terminal carbon atom. The second isomer results from replacement of a hydrogen atom bonded to either center carbon atom

CH_3—CH_2—CH_2—CH_2—Br

(I)

CH_3—CH_2—CH—CH_3
            |
            Br

(II)

The brominated product has only two isomers, since the parent hydrocarbon butane has only two types of equivalent hydrogen atoms. Note that the alternative representations (III) and (IV) are not additional isomers.

Br—CH_2—CH_2—CH_2—CH_3

(III)

CH_3—CH—CH_2—CH_3
       |
       Br

(IV)

They result from substituting bromine on the second carbon atom of each of the two sets of equivalent carbon atoms. (III) is the same isomer as (I), and (IV) is the same isomer as (II).

A systematic method of writing all chain isomers for a given hydrocarbon can be developed by using the equivalency of carbon atoms. The first isomer of an alkane having *n* carbon atoms is a continuous chain of *n* carbon atoms. Then a chain of (*n* – 1) carbon atoms is written, and the number of equivalent carbon atoms is determined. The remaining carbon atom may be bonded to any of the nonequivalent carbon atoms (other than the end carbon atoms) and from each a different isomeric structure results. Then a chain of (*n* – 2) carbon atoms is written, and the number of equivalent carbon atoms is determined. The remaining two carbon atoms must then be placed so that isomeric structures result. This is done by placing one carbon atom in each of the possible nonequivalent positions, reexamining for equivalence in the resulting structure, and then adding the remaining carbon atom.

The isomers of molecular formula C_5H_{12} are shown in Figure 24-3. The structure (a) has an extended chain of five carbon atoms.

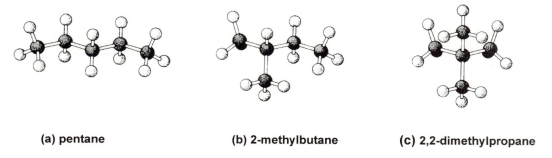

| (a) pentane | (b) 2-methylbutane | (c) 2,2-dimethylpropane |

Figure 24-3. Isomeric pentanes.

Next consider a structure of four carbon atoms—that is, butane—and determine where the fifth carbon atom necessary to give C_5H_{12} may be placed. Butane has two sets of equivalent carbon atoms. However, the attachment of the fifth carbon atom at a terminal carbon atom produces a chain of five carbon atoms, which has already been considered. Thus, the fifth carbon atom must be placed at an interior carbon atom to give structure (b).

Next consider a chain of three carbon atoms. The remaining two carbon atoms can be placed only on the interior carbon atom as shown in structure (c). If either or both of the remaining two carbon atoms were placed on a terminal carbon atom, a longer chain of atoms would result. Such structures have already been constructed.

PROCEDURE

Methane

Construct two models of CH_4. Place them on the desk and observe that three hydrogen atoms are in contact with the desk forming a tripod. The remaining hydrogen atom is pointed up. Orient the models so that one of the hydrogen atoms of the tripod is pointed to the right. One of the remaining hydrogen atoms will be directed to the left and the front of the desk, whereas the other will be directed to the left and the back of the desk.

Place a piece of masking tape on the "right" hydrogen atom on one model and a piece of masking tape on the "left and front" hydrogen atom on the other model. Keeping the tripod in contact with the desk, rotate one of the models so that the taped atom of model one is in the same relative position as the taped atom in model two. What does this experiment show about the equivalence of the hydrogen atoms forming the tripod?

Remove the masking tape from one of the models and place it on the hydrogen atom pointing up. Now pick up this model and reposition it next to the other model so that both taped hydrogen atoms are identically oriented on the table top. What does this show about the equivalence of the hydrogen atoms in methane? Write a two-dimensional structure of methane on your Report Sheet.

Molecular Models

Remove the tape from the hydrogen atoms of the two models. Replace the hydrogen atom pointing up in one model by a bromine atom. Replace one of the hydrogen atoms of the tripod in the other model by a bromine atom. Now convince yourself that these structures represent the same molecule. Write the structure of CH_3Br on your Report Sheet.

Place both models with the bromine atom pointed up. Examine the three hydrogen atoms in the tripod and tape different positions on each of the two models. Now rotate one of the models until the taped hydrogen atoms occupy identical positions. When the taped atoms are replaced by a bromine atom, do isomers result? Write the CH_2Br_2 structure on your Report Sheet.

Successively replace the two remaining hydrogen atoms to obtain $CHBr_3$ and CBr_4. Draw structural formulas for these compounds on your Report Sheet.

Ethane

Construct two models of ethane and arrange them in the staggered and eclipsed conformations shown in Figure 24-4. By rotation about the carbon-carbon bond the staggered conformation on the left can be converted into the eclipsed conformation on the right.

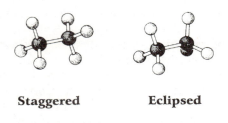

Staggered   **Eclipsed**

Figure 24-4. Conformations of ethane.

Now tape the "right up" hydrogen atom in one eclipsed conformation; in the other model tape one of the two hydrogen atoms on the right carbon in contact with the desk. By picking up and reorienting one of the models, but without rotating the carbon–carbon bond, show that the two models are equivalent. Now reorient the models to their original positions and show that one can be converted to the other by rotation about the carbon–carbon bond. What do these experiments indicate about the equivalence of the three hydrogen atoms on the right carbon atom?

Now remove the tape from one of the models and place it on the "left up" hydrogen atom. Orient the other model so that the tape is on the "right up" hydrogen atom. Is there a way of reorienting one of these models so that it is equivalent to the other model? What does this experiment show about the equivalence of the hydrogen atoms on the two carbon atoms? Write the structural formula for ethane on your Report Sheet.

Now orient the taped hydrogen atoms in both models to the "right up" position. Replace these hydrogen atoms with bromine atoms. Would different models have been produced if untaped hydrogen atoms had been replaced?

Examine the carbon atoms in the model of CH_3CH_2Br. Are they equivalent? In one model replace another hydrogen atom on the right carbon atom with a bromine atom. In the second model use a bromine atom to replace a hydrogen atom on the left carbon atom. Is it possible to reorient these models or to rotate the carbon-carbon bonds so that the two models become equivalent? What does this test indicate about the number of isomers of $C_2H_4Br_2$?

Draw structural formulas for the isomers of $C_2H_4Br_2$ on your Report Sheet. Name these compounds.

Propane

Construct a model for C_3H_8 by replacing one of the hydrogen atoms of ethane by a $-CH_3$ unit. Examine the model and determine whether the two terminal carbon atoms are equivalent or nonequivalent. Are the hydrogen atoms on the two terminal carbon atoms equivalent or nonequivalent?

Examine the central carbon atom. Are the two attached hydrogen atoms equivalent to the hydrogen atoms on the terminal carbon atoms?

How many isomers of C_3H_7Br can result by replacing a hydrogen atom of C_3H_8 with a bromine atom? Construct each possible model. Write the structural formula for each isomer on your Report Sheet and name it.

Examine the model containing a bromine atom attached to the interior carbon. How many equivalent carbon atoms are in this model? How many different models can result by replacing different hydrogen atoms by a second bromine atom? Write structural formulas for the isomers on your Report Sheet and name them.

Examine the model containing a bromine atom attached to a terminal carbon atom. How many types of equivalent carbon atoms are in this model? How many different models can result by replacing different hydrogen atoms by a second bromine atom? Prepare these models and write structural formulas for them on your Report Sheet. Are any of these models the same as any prepared by starting with the first bromine on an interior carbon atom? How many isomers are there with the molecular formula $C_3H_6Br_2$? Name them.

The molecular formula $C_3H_5Br_3$ has five isomers. Devise a method of preparing a model of each isomer and write its corresponding structural formula on your Report Sheet. Name each isomer.

Butanes

Prepare a model for butane by adding a $-CH_3$ unit to the terminal carbon atom of propane in place of a hydrogen atom. Examine each carbon atom and its attached hydrogen atoms. Convince yourself that only two types of nonequivalent carbon atoms exist.

Prepare the isomer of butane called 2-methylpropane by adding a $-CH_3$ unit to the interior carbon atom of propane in place of a hydrogen atom. Examine each carbon atom and convince yourself that one carbon atom is unique. Are the other three carbon atoms equivalent or nonequivalent?

How many isomers of the formula C_4H_9Br can be constructed from butane by the replacement of a hydrogen atom? How many isomers of formula C_4H_9Br can be constructed from 2-methyl-propane by the replacement of a hydrogen atom? Write structural formulas for all C_4H_9Br isomers on your Report Sheet. Name them.

Starting from butane write the structural formulas for six isomers with the molecular formula $C_4H_8Br_2$ on your Report Sheet. Name them.

Do likewise starting from 2-methylpropane and write the structural formulas for three additional isomers with the molecular formula $C_4H_8Br_2$ on your Report Sheet. Name them.

Pentanes

Examine the three isomers of C_5H_{12} in Figure 24-3. How many sets of equivalent carbon atoms are in each compound? Using this information, write the structural formulas for eight isomeric compounds with the molecular formula $C_5H_{11}Br$ on your Report Sheet. Name them.

Molecular Models

Hexanes

Prepare models of the five isomeric compounds with the molecular formula C_6H_{14}. Write your structural formulas on your Report Sheet.

Report Sheet

_____ _____ _____
Name   Instructor/Section   Date

Methane

Compound	Number of equivalent hydrogen atoms
CH_4	_____
CH_3Br	_____
CH_2Br_2	_____
$CHBr_3$	_____
CBr_4	_____

Draw the structural formulas for CH_4, CH_3Br, CH_2Br_2, $CHBr_3$, CBr_4 .

CH_4 CH_3Br CH_2Br_2 $CHBr_3$ CBr_4

Ethane

How many equivalent hydrogen atoms are there in C_2H_6? _____

How many sets of equivalent carbon atoms are there in ethane? _____

Draw structural formulas for C_2H_5Br, the two isomeric $C_2H_4Br_2$ compounds, and the two isomeric $C_2H_3Br_3$ compounds. Name each isomer.

C_2H_5Br $C_2H_4Br_2$ $C_2H_4Br_2$ $C_2H_3Br_3$ $C_2H_3Br_3$

189

Propane

How many sets of equivalent carbon atoms are there in propane? Write the structural formulas for the two isomeric C_3H_7Br compounds. Name each compound.

How many sets of equivalent carbon atoms are there in each isomer of C_3H_7Br? Explain.

Draw structural formulas for the isomeric $C_3H_6Br_2$ compounds and name them. How many sets of equivalent carbon atoms are there in each isomer?

Draw structural formulas for the isomeric $C_3H_5Br_3$ compounds and name them. How many sets of equivalent carbon atoms are there in each isomer?

Butanes

Identify each set of equivalent carbon atoms in butane and count the number of equivalent carbon atoms in each set. Do the same for 2-methylpropane.

Draw structural formulas for the four isomeric compounds with the molecular formula C_4H_9Br. Name each isomer. How many sets of equivalent carbon atoms are there in each isomer?

Draw the structural formulas for the nine isomers with the molecular formula $C_4H_8Br_2$. Name each isomer.

Molecular Models

Pentanes

Draw the structural formulas for the eight isomeric compounds with the molecular formula $C_5H_{11}Br$. Name each isomer.

Hexanes

Write the structural formulas for the five isomeric compounds with the molecular formula C_6H_{14}. How many sets of equivalent carbon atoms are there in each isomer? Name each isomer.

EXPERIMENT

25   Hydrocarbons

APPARATUS

10 × 75 mm test tubes, 16 × 150 mm test tubes, acetylene generator, medicine dropper, watch glass, test tube rack.

REAGENTS

Heptane, 1-octene, cumene, 1-butanol, ligroin, 1,4-dioxane, 1% bromine in 1,4-dioxane, 1% aqueous potassium permanganate, unknown samples of hydrocarbons, calcium carbide.

INTRODUCTION

The hydrocarbons are a large class of compounds, but they can be divided into subclasses based on structural features. These structural features are also associated with common chemical reactivities. Thus, each subclass is distinguished by both structure and reactivity. In this experiment you will examine the chemical reactivity of the saturated, unsaturated, and aromatic hydrocarbons.

Structures of Hydrocarbons

A hydrocarbon is saturated if it has the maximum possible number of hydrogen atoms bonded to its carbon atoms. This occurs only if the carbon atoms are bonded to each other only by single bonds known as sigma (σ) bonds. Ethane is a saturated hydrocarbon based on this criterion.

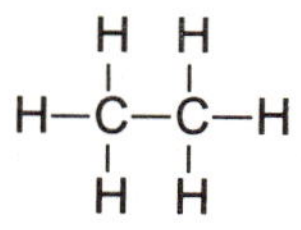

$$
\begin{array}{ccc}
& H & H \\
& | & | \\
H- & C- & C-H \\
& | & | \\
& H & H
\end{array}
$$

Ethane

Unsaturated hydrocarbons have fewer than the maximum number of hydrogen atoms. Such compounds must have carbon-carbon multiple bonds so that the total number of covalent bonds to each carbon atom is maintained at four. Ethylene, acetylene, and benzene are unsaturated hydrocarbons.

Ethylene   Acetylene   Benzene

The two carbon-carbon bonds in ethylene consist of one σ bond similar to that in saturated hydrocarbons and a second type of bond called a pi (π) bond. Together they form a double bond. Acetylene has a triple bond consisting of one σ bond and two π bonds.

Aromatic hydrocarbons, such as benzene, have a unique bond between carbon atoms that is not easily represented. In benzene, an aromatic hydrocarbon, all six carbon-carbon bonds are identical. Two representations of benzene, that are resonance contributors, have a series of alternating single and double bonds which are used to represent the structure of this aromatic hydrocarbon.

These representations indicate that there are six electrons equally distributed in an aromatic π network over all six carbon atoms. The bonds represented by the sides of the hexagon are σ bonds. The actual structure of benzene is a resonance hybrid of the two structures shown.

Reactivity of Hydrocarbons

The σ bonds of saturated hydrocarbons (alkanes) are very stable and hence very unreactive. At high temperatures, saturated hydrocarbons react with oxygen (they burn). In this reaction carbon-carbon bonds are broken and the products are carbon dioxide and water. If the combustion is inefficient, some carbon monoxide and even free carbon (soot) may be formed. In general, saturated hydrocarbons burn more efficiently than the other classes of hydrocarbons.

The carbon-hydrogen bonds of alkanes can be replaced or substituted by halogens while maintaining the carbon framework. A general equation for the reaction with bromine is

$$R\!-\!H + Br_2 \longrightarrow R\!-\!Br + HBr$$

Note that HBr is a product of the reaction. The reaction with halogens requires either heat or light energy.

The π bonds of unsaturated hydrocarbons (alkenes and alkynes) are reactive and undergo addition reactions. In these reactions a molecule such as bromine forms two carbon-bromine single bonds at the expense of the π bond. Ethylene adds bromine to form 1,2-dibromoethane.

(colorless)   (red-brown)   (colorless)

The reaction occurs instantly at room temperature. As a result the characteristic red-brown color of bromine disappears. Note also that no HBr is formed in this reaction as in the substitution reaction of saturated hydrocarbons. Acetylene, which has two π bonds, also undergoes addition reactions.

$$H-C\equiv C-H \ + \ 2\ Br_2 \longrightarrow \ \underset{\underset{Br\ \ Br}{|\quad|}}{\overset{\overset{Br\ \ Br}{|\quad|}}{H-C-C-H}}$$

   (colorless)   (red-brown)   (colorless)

The π bond is also a center for attack by oxidizing agents. For example, dilute neutral potassium permanganate solutions react with alkenes and alkynes to produce dialcohols. Visually the reaction is accompanied by the loss of the purple color of potassium permanganate and the formation of a brown precipitate of manganese dioxide. The application of this reaction to test for unsaturation is called the Baeyer test.

$$3\ RCH{=}CHR \ + \ 2\ KMnO_4 \ + \ 4\ H_2O \longrightarrow \ 3\ \underset{\underset{OH\ \ OH}{|\quad\ |}}{\overset{\overset{OH\ \ OH}{|\quad\ |}}{RCH-CHR}} \ + \ 2\ MnO_2(s) \ + \ 2\ KOH$$

               (purple)   (brown)

In strongly basic solution, the purple permanganate ion reacts to form a soluble green product instead of the solid brown oxide. This green species is the manganate (MnO_4^{2-}) ion.

The π network of aromatic compounds is maintained in their reactions. Saturated groups attached to the benzene ring can be attacked by vigorous oxidizing agents. The product is always benzoic acid when any single alkyl group is attached to the benzene ring. Additional alkyl groups would also be oxidized to $-CO_2H$.

PROCEDURE

Before performing this experiment study Section A, "Handling Liquid Reagents," on pages 6-7 of this manual. Always discard organic waste into the special can provided.

    Caution: Never pour organic liquids (hydrocarbons) in the sink.

Use a clean, dry test tube for each test. Use heptane, 1-octene, and cumene. The structures of the hydrocarbons are

     $CH_3(CH_2)_5CH_3$ $CH_3(CH_2)_5CH{=}CH_2$

       Heptane   1-Octene   Cumene

Hydrocarbons

Many commercial alkanes contain alkenes as impurities. To extract these unsaturated impurities completely, the alkane is treated with concentrated sulfuric acid. To three parts of alkane is added one part H_2SO_4. The lower (H_2SO_4) layer darkens and is removed. The remaining alkane layer is treated again with H_2SO_4 until it no longer darkens. The alkane then is washed with water.

CAUTION: Because of the hazards of concentrated H_2SO_4, this should be performed only by experienced personnel.

Part A—Solubility of Hydrocarbons

Test the solubility of hydrocarbons in the following solvents by shaking 0.5 mL of each compound in a 16 × 150 mm test tube with 5 mL of each solvent—water, 1-butanol, and ligroin (a mixture of alkanes). Record your observations on your Report Sheet.

Part B—Flammability of Hydrocarbons

CAUTION: Hydrocarbons are highly flammable, and their vapors may be explosive in air. Be very careful with flames. Do not use any more than the small quantities of hydrocarbons indicated.

Place 3 drops of one of the hydrocarbons on a watch glass and, using a match, ignite the sample. Observe the type and color of the flame, the carbon that is in the flame, and the amount of residue that remains. Repeat this experiment with each of the other hydrocarbons. Record your observations on your Report Sheet

Part C—Action of Bromine in 1,4-Dioxane on Hydrocarbons

Place 1 mL of 1% bromine in 1,4-dioxane in a micro test tube. Add 10-20 drops of a hydrocarbon and note any color changes that result. Perform this test on each hydrocarbon. Record your observations on your Report Sheet.

Part D—Reaction of Hydrocarbons with Potassium Permanganate

CAUTION: The 6 M solution of sodium hydroxide is caustic. Rinse your skin with water immediately if it has a soapy feeling. Sodium hydroxide will cause irreversible eye damage and blindness. Do not touch your eyes with your hands if you have spilled sodium hydroxide on your hands.

Place 1 mL of a hydrocarbon in a micro test tube and then add 5 drops of a 1% aqueous solution of potassium permanganate. Then add 1 drop of 6 M sodium hydroxide. Shake the tube and observe all the changes that occur. How much time elapses before a change occurs? Wait 5 minutes before recording your final observations.

Part E—Classification of an Unknown

Obtain an unknown hydrocarbon sample from your laboratory instructor and run selected tests from Parts A through D to classify it as saturated, unsaturated, or aromatic.

Part F—Preparation and Chemical Properties of Acetylene

Your instructor will arrange an acetylene generator in the hood. It consists of a small, dry 250 mL bottle fitted with a two-hole rubber stopper, a dropping funnel, glass tubing, and rubber tubing to fit (Figure 25-1). Several small pieces of calcium carbide (CaC_2) are placed in the dry bottle. Acetylene gas is then generated by the slow, cautious addition of water from the dropping funnel. To a dry 16 × 150 mm test tube containing 5 drops of a 1% solution of bromine in 1,4-dioxane, add about 3 mL of 1,4-dioxane. Bubble acetylene through this solution, and observe any change in the color of the solution in the test tube. Record your observations on your Report Sheet.

To a 16 × 150 mm test tube containing 2 drops of a 1% aqueous solution of potassium permanganate, add about 3 mL of water. Bubble acetylene through this solution until the permanganate color disappears. If the reaction is slow, shake and continue the bubbling. Repeat this procedure if there is no decolorization. Record your observations on your Report Sheet.

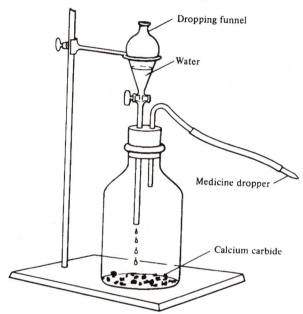

Figure 25-1. Acetylene Generator.

Report Sheet

Name		Instructor/Section	Date

Part A—Solubility of Hydrocarbons

Hydrocarbon	Water	1-Butanol	Ligroin
Heptane	_____	_____	_____
1-Octene	_____	_____	_____
Cumene	_____	_____	_____

Part B—Flammability of Hydrocarbons

Hydrocarbon	Color and type of flame	Amount of residue
Heptane	_____	_____
1-Octene	_____	_____
Cumene	_____	_____

Part C—Action of Bromine on Hydrocarbons

Reactants	Observations	Equation (if a reaction occurs)
Heptane with Br_2	_____	_____
1-Octene with Br_2	_____	_____
Cumene with Br_2	_____	_____

Part D—Reaction of Hydrocarbons with Potassium Permanganate in Basic Solution

Hydrocarbon	Observations	Conclusions
Heptane	_____	_____
1-Octene	_____	_____
Cumene	_____	_____

Part E—Classification of an Unknown

Unknown No. _____

Reagent	Observations	Conclusions
_____	_____	_____
_____	_____	_____
_____	_____	_____

Part F—Preparation and Chemical Properties of Acetylene

Reagent	Observations	Conclusions
Br_2	_____	_____
$KMnO_4$	_____	_____

Answer the questions on the following pages.

QUESTIONS

1. Give the expected products of each of the following reactions.

a. $CH_3CH_2C\equiv CCH_3$ + O_2 $\xrightarrow{\text{heat}}$

b. $CH_3CH_2CH=CHCH_3$ + O_2 $\xrightarrow{\text{heat}}$

c. 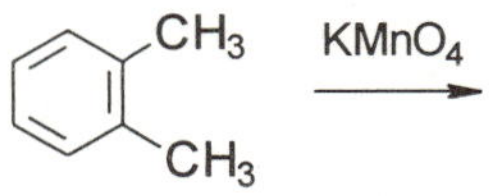 $\xrightarrow{\text{KMnO}_4}$

d. 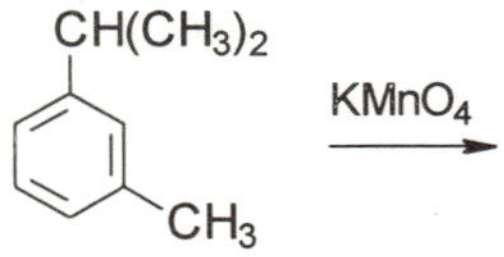 $\xrightarrow{\text{KMnO}_4}$

e. 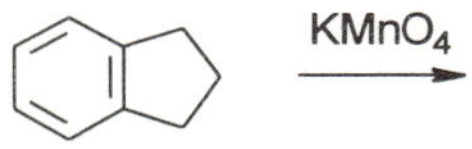 $\xrightarrow{\text{KMnO}_4}$

f. 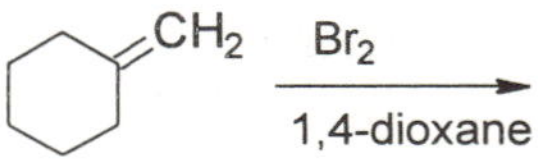 $\xrightarrow{\text{Br}_2}{\text{1,4-dioxane}}$

g. 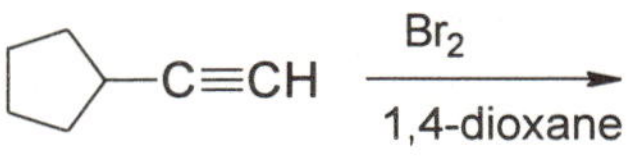 $-C\equiv CH$ $\xrightarrow[\text{1,4-dioxane}]{\text{Br}_2}$

h. $CH_3CH_2CH=CHCH_2CH=CH_2$ $\xrightarrow[\text{1,4-dioxane}]{\text{Br}_2}$

i. $CH_3CH_2CH{=}CHCH_3$ $\xrightarrow{\text{KMnO}_4}$

j. $\xrightarrow[\text{1,4-dioxane}]{\text{Br}_2}$

k. CaC_2 + H_2O $\longrightarrow$

EXPERIMENT

26 Properties of *Cis-Trans* Isomers

APPARATUS

125 mL Erlenmeyer flask, plastic air condenser, microspatula, thermometer, capillary tubes, filter paper, eight 10 × 75 mm test tubes, watch glass, ring stand, iron ring, clamp, wire gauze, burner, rubber tubing, matches, buret clamp, melting temperature apparatus.

REAGENTS

Maleic acid, concentrated hydrochloric acid, magnesium ribbon (3 cm strips), sodium carbonate, thymol blue, pH paper.

INTRODUCTION

Maleic acid and fumaric acid have the same molecular formula: $C_4H_4O_4$. Each compound contains two carboxyl groups, –COOH, and a double bond. The compounds are isomers because their structures differ in the geometry, or arrangement in space, of the atoms in the molecule. Examine the structural formulas of the two acids shown below and note that in the *trans* form the carboxyl groups are on opposite sides of the molecule and in the *cis* form they are on the same side. The fact that the two isomers can be isolated indicates that rotation of the molecule at the double bond is restricted.

cis isomer   *trans* isomer

As a result of this geometrical difference, the *cis* isomer and the *trans* isomer have different physical properties. In addition, some of their chemical properties differ because of the locations of the two carboxyl groups.

In this experiment you will convert maleic acid to fumaric acid by heating it in an aqueous solution containing some hydrochloric acid (HCl). The HCl is not used up in the reaction but serves as a source of $H^+(aq)$, which is the catalyst. You will compare some of the properties of each acid and attempt to explain any differences in terms of structures. This should enable you to draw some conclusions about which acid, maleic or fumaric, is the *trans* form and which is the *cis* form.

PROCEDURE

CAUTION: Concentrated hydrochloric acid (HCl) causes severe skin burns. Immediately wash the affected areas with cold running water and promptly report the accident to your instructor.

Part A—Conversion of Maleic Acid to Fumaric Acid

Weigh 4.0 g of maleic acid in a clean, dry 125 mL Erlenmeyer flask. Add 8 mL of distilled water and warm slightly to dissolve the acid. Add 10 mL of concentrated HCl and insert a plastic air condenser (see Figure 26-1).

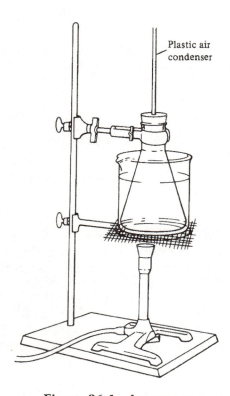

Plastic air condenser

Figure 26-1. Apparatus assembly.

Place the flask in a hot water bath assembly as shown in the figure, and clamp the flask in place under the hood. Heat until solid fumaric acid forms in the flask (approximately 10 minutes). Cool the solution to room temperature by slowly placing the Erlenmeyer flask into a cold water bath. After the solution has cooled, collect the crystals of fumaric acid using vacuum filtration as described in Section D of Laboratory Techniques on page 10. Rinse any remaining crystals in the Erlenmeyer flask into the Büchner funnel using approximately 10 mL of cold distilled water. Take a small sample of fumaric acid (about the size of 2 grains of rice) and allow it to air dry until the next laboratory period when the melting point determination is done for Part B.

Part B—Comparison of the Two Isomers

Solubility. Compare the solubility of the two acids by placing one-half microspatulaful of each into separate properly labeled 10 × 75 mm test tubes. Add 2 mL of distilled water to each and make a qualitative comparison of solubility. Record your observations on the Report Sheet.

Chemical Properties. Compare the chemical properties of the two isomers as follows. Prepare four solutions of maleic acid by adding approximately one-quarter microspatulaful to each of four 10 × 75 mm micro test tubes. Add 2 mL of distilled water to each. Prepare four solutions of fumaric acid in the same way.

Perform each of the following tests on one solution each of fumaric acid and maleic acid. After each test, compare and record the results on the Report Sheet before continuing.
1. Estimate the hydrogen ion concentration by testing one solution of each acid with pH paper.
2. To another solution of each acid, add a 3 cm strip of magnesium ribbon.
3. To the third solution of each acid, add a small amount of sodium carbonate.
4. To the fourth solution of each acid, add 2 drops of thymol blue indicator.

Melting Points. Study Section K, "Melting Point," on pages 22-24. Obtain two capillary melting tubes. Make one tube shorter than the other by cutting off approximately 1 centimeter. Place maleic acid in the shorter tube to identify it. Place fumaric acid in the longer tube. For the compound whose melting point is to be determined, use a spatula to crush a quantity about the size of a small grain of rice on a watch glass. Grind the particles small enough to fit into the melting-point capillary. Use the open end of the capillary to pick up a portion of the compound, and carefully tap the capillary with your finger in order to vibrate the sample to the bottom of the tube. Repeat until the compound in the bottom of the tube attains a depth of about 2 mm.

If the approximate melting point is not known, time can be saved by first doing a rough determination. For the rough determination, the sample is heated rapidly until the solid melts. Then a second precise determination on a new sample is done. Allow the melting-point apparatus to cool to a temperature 5-10 degrees lower than the roughly determined melting point. Then continue to heat very slowly while watching the sample in the capillary tube. Record two temperatures, the temperature at which melting first starts and the temperature at which the last portion of solid has just melted. This is the melting range of the solid. Only one of the samples will melt below 180°C. Stop heating when one of the samples has melted. The other sample melts at 287°C.

Report Sheet

Name		Instructor/Section	Date

Comparison of the two isomers

Property	Maleic acid	Fumaric acid
Solubility	_____	_____
Melting point	_____	_____
pH of aqueous solution	_____	_____
Reaction with magnesium ribbon	_____	_____
Reaction with sodium carbonate	_____	_____
Reaction with Thymol Blue	_____	_____

Answer the questions on the next page.

QUESTIONS

1. Assume that equilibrium concentrations were achieved in Part A. Which acid is the most stable?

2. Some diacids can lose a molecule of water when the two carboxyl groups react to form a cyclic anhydride. Phthalic acid is a diacid that reacts as follows.

Maleic acid also can lose a molecule of water and form maleic anhydride. Which structural isomer, *cis* or *trans*, do you predict it is? Fumaric acid cannot do this. Explain.

3. What do each of the following experiments contribute to your knowledge of the structure of each isomer?

   a. The melting point determination.

   b. The reactions of solutions of each acid with the indicator, thymol blue.

   c. The reactions of each with sodium carbonate.

   d. The reactions of each with magnesium.

EXPERIMENT

27   Alcohols

APPARATUS

10 × 75 mm test tubes, 16 × 150 mm test tubes, beakers, medicine dropper, test tube rack.

REAGENTS

Methanol, ethanol, 1-propanol, 2-propanol, 1-butanol, 2-butanol, 2-methyl-2-propanol, 1-pentanol, 1-octanol, 10% aqueous sodium dichromate ($Na_2Cr_2O_7$), 3 M NaOH, 6 M HCl, iodine-potassium iodide solution, Lucas reagent.

INTRODUCTION

Alcohols can be considered organic analogs of water. In an alcohol, the hydroxyl group (–OH) is attached to a saturated carbon atom rather than to a hydrogen atom as in water. The presence of the hydroxyl group results in physical and chemical properties distinctly different from those of hydrocarbons. The structural features of the hydroxyl group that are responsible for the characteristic properties of alcohols are the polar hydrogen-oxygen bond and the lone-pair electrons of the oxygen atom.

Some differences in the chemistry of alcohols are based on the structure of the alkyl group. One important feature is the number of carbon atoms bonded to the –C–OH group. Primary, secondary, or tertiary alcohols have 1, 2, or 3 carbon groups, respectively, bonded to the carbon atom bonded to oxygen.

$$
\begin{array}{ccc}
\text{H} & \text{H} & \text{R} \\
| & | & | \\
\text{R–C–OH} & \text{R–C–OH} & \text{R–C–OH} \\
| & | & | \\
\text{H} & \text{R} & \text{R} \\
\text{Primary alcohol} & \text{Secondary alcohol} & \text{Tertiary alcohol}
\end{array}
$$

The hydroxyl group of alcohols can form hydrogen bonds with water. As a consequence, alcohols are more soluble in water than hydrocarbons. While the lower molecular weight alcohols are miscible in all proportions with water, the solubility of higher molecular weight alcohols approaches that of the alkanes. As the size of the alkyl group increases, the polar hydroxyl group becomes a smaller contributor to the overall physical properties of the alcohol.

The proton of the hydroxyl group is less acidic than the proton of water. Therefore, the proton of the hydroxyl group can be removed only by strong bases.

209

Alcohols

The proton of the hydroxyl group of alcohols can be reduced by active metals to produce hydrogen gas. This reaction parallels that of active metals with water.

$$2\ ROH + 2\ Na \longrightarrow 2\ RONa + H_2$$
$$2\ HOH + 2\ Na \longrightarrow 2\ NaOH + H_2$$

The rate of this reaction is related to the acidity of the alcohols. The most acidic compounds are the most reactive toward sodium.

The hydroxyl group of alcohols may be replaced by other groups such as halogens. The rate of such reactions depends on the structure of the alkyl group. Replacement of the hydroxyl group by a chloro group can be accomplished by a variety of reagents. One such reagent, consisting of $ZnCl_2$ and HCl, is the Lucas reagent.

$$ROH + HCl \xrightarrow{ZnCl_2} RCl + H_2O$$

In tertiary alcohols, the reaction occurs nearly instantly at room temperature. Secondary alcohols react within 5-10 minutes. Primary alcohols will react only after hours at room temperature. Since the alkyl chloride is insoluble in the Lucas reagent, a cloudiness in the solution indicates that a reaction has occurred. The time required for the onset of turbidity is indicative of the structure of the alcohol.

Primary, secondary, and tertiary alcohols behave in characteristic ways toward oxidizing agents. One common oxidizing agent is sodium dichromate ($Na_2Cr_2O_7$) in dilute hydrochloric acid. Primary alcohols are oxidized to aldehydes, which in turn are oxidized to acids. Secondary alcohols are oxidized to ketones, which are not further oxidized. Tertiary alcohols are not oxidized by sodium dichromate in dilute hydrochloric acid.

$$RCH_2OH \xrightarrow[HCl]{Na_2Cr_2O_7} RCHO \xrightarrow[HCl]{Na_2Cr_2O_7} RCO_2H$$

$$\underset{RCHR}{\overset{OH}{|}} \xrightarrow[HCl]{Na_2Cr_2O_7} \underset{RCR}{\overset{O}{\|}}$$

$$R_3COH \xrightarrow[HCl]{Na_2Cr_2O_7} \text{no reaction}$$

The chromium in dichromate ion is reduced to Cr^{3+} in the reaction. Solutions of dichromate ion are yellow-orange, whereas solutions of Cr^{3+} are green. Thus, oxidation of an alcohol by sodium dichromate is indicated by the color change of the chromium species.

Alcohols with the following partial structure react in a series of reactions with iodine and sodium hydroxide to produce iodoform, a yellow solid.

$$\underset{H}{\overset{OH}{\underset{|}{R-\underset{|}{C}-CH_3}}} \xrightarrow[NaOH]{I_2} \overset{O}{\overset{\|}{R-C-CH_3}} \xrightarrow[NaOH]{I_2} \overset{O}{\overset{\|}{R-C-Cl_3}} \xrightarrow[NaOH]{I_2} \overset{O}{\overset{\|}{R-C-O^-}} + CHI_3$$

iodoform
(yellow precipitate)

This reaction can be useful in characterizing the structure of an alcohol.

PROCEDURE

Part A—Solubility of Alcohols in Water

Determine the solubilities of ethanol, 1-butanol, 2-butanol, 2-methyl-2-propanol, 1-pentanol, and 1-octanol. Place 1 mL of water in a micro test tube. Add the sample of an alcohol dropwise with shaking, until it no longer dissolves or until the volume of the solution is double that of the water originally present. Repeat this procedure for each alcohol. Characterize the alcohols as insoluble, slightly soluble, or completely soluble on your Report Sheet.

Part B—Reactivity of Alcohols Toward Sodium

Your laboratory instructor will demonstrate the reaction of metallic sodium with methanol, ethanol, 1-propanol, 2-propanol, and 2-methyl-2-propanol. Record your observations on your Report Sheet, and write chemical equations for the reactions. Arrange these alcohols in order of increased acidity of the O—H bond.

Part C—Lucas Test

CAUTION: The Lucas reagent contains concentrated hydrochloric acid (HCl)which causes severe skin burns. Immediately wash the affected areas with cold running water and promptly report the accident to your instructor.

Determine the reactivities of 1-butanol, 2-butanol, and 2-methyl-2-propanol. Place 1 mL of each alcohol in a 16 × 150 mm test tube and add 5 mL of Lucas reagent. Stopper the tubes, mix the contents well by shaking, and allow the tubes to stand at room temperature. Observe what happens immediately, after 5 minutes, and again after 30 minutes. Record your observations on the Report Sheet, and write equations for all of the reactions. If no reaction took place, write "N.R."

Part D—Oxidation of Alcohols

Perform this test with 1-butanol, 2-butanol, and 2-methyl-2-propanol. Place 1 mL of the alcohol to be tested and 10 drops of 6 M HCl in a micro test tube. Add 1 drop of 10% $Na_2Cr_2O_7$ solution and shake. Observe and record any color changes on your Report Sheet.

Part E—Iodoform Reaction

Perform this test with 1-propanol, 2-propanol, 2-butanol, and 2-methyl-2-propanol. Dissolve 4 drops of the compound to be tested in 1 mL of methanol and 1 mL of water. Then add 2 mL of 3 M NaOH solution. Add I_2–KI solution dropwise until a definite yellow color persists for at least 2 minutes after mixing. Allow the test tube to stand for 5 minutes, and record your observations on the Report Sheet.

Report Sheet

Part A—Solubility of Alcohols in Water

Compound	Observations
Ethanol	
1-Butanol	
2-Butanol	
2-Methyl-2-propanol	
1-Pentanol	
1-Octanol	

Summarize your observations on the solubility of alcohols.

Part B—Reactivity of Alcohols Toward Sodium

Compound	Observations
Methanol	
Ethanol	
1-Propanol	
2-Propanol	
2-Methyl-2-propanol	

Summarize how the reactivity of alcohols depends on the structure of the alkyl group.

Alcohols

Part C—Lucas Test

Compound	Observations	Equation
1-Butanol	_____	_____
2-Butanol	_____	_____
2-Methyl-2-propanol	_____	_____

Summarize the reactivities of alcohols toward the Lucas reagent.

Part D—Oxidation of Alcohols by Acidic $Na_2Cr_2O_7$ Solution

Compound	Observations	Equation
1-Butanol	_____	_____
2-Butanol	_____	_____
2-Methyl-2-propanol	_____	_____

Summarize the reactivities of alcohols toward oxidizing agents.

Part E—Iodoform Test

Compound	Observations	Equation
1-Propanol	_____	_____
2-Propanol	_____	_____
2-Butanol	_____	_____
2-Methyl-2-propanol	_____	_____

Summarize the reactivities of alcohols toward I_2 and NaOH.

Answer the questions on the following pages.

QUESTIONS

1. A compound of formula $C_4H_{10}O$ does not react with sodium, Lucas reagent, or sodium dichromate solution. Write three possible structures for this compound. Explain their lack of reactivity.

2. *trans*-1,4-Cyclohexanediol is very soluble in water, whereas *cis*-1,2-dimethoxycyclobutane is not. Explain why this difference is observed for these isomers.

3. There are 17 isomeric alcohols of formula $C_6H_{14}O$. Only four of these isomers give a positive iodoform test. Draw these four structures.

4. There are 17 isomeric alcohols of formula $C_6H_{14}O$. Three isomers react rapidly with the Lucas reagent. Write the structures of these alcohols.

5. What are the products of each of the following reactions? If no reaction occurs, write "N.R."

a. 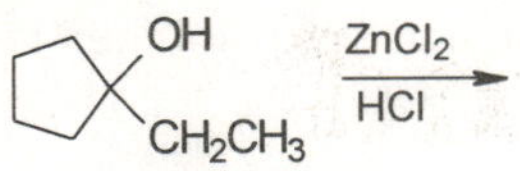
$$\text{1-ethylcyclopentan-1-ol (OH, CH}_2\text{CH}_3\text{)} \xrightarrow[\text{HCl}]{\text{ZnCl}_2}$$

b. 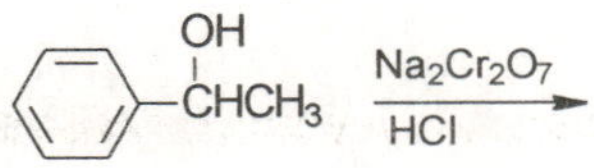
$$\text{(phenyl)CHCH}_3\text{(OH)} \xrightarrow[\text{HCl}]{\text{Na}_2\text{Cr}_2\text{O}_7}$$

c. 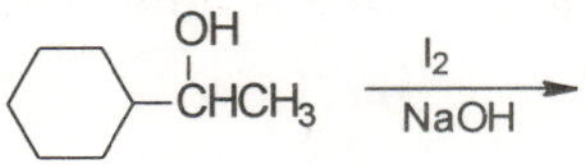
$$\text{(cyclohexyl)CHCH}_3\text{(OH)} \xrightarrow[\text{NaOH}]{\text{I}_2}$$

d. 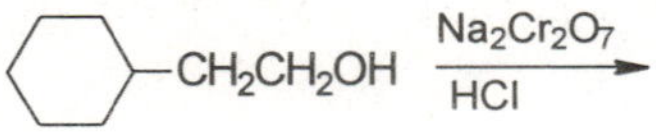
$$\text{(cyclohexyl)CH}_2\text{CH}_2\text{OH} \xrightarrow[\text{HCl}]{\text{Na}_2\text{Cr}_2\text{O}_7}$$

e. 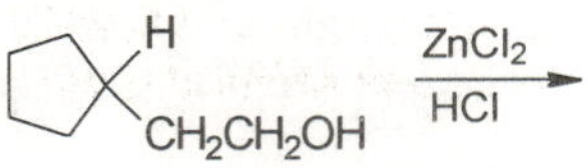
$$\text{(cyclopentyl with H and CH}_2\text{CH}_2\text{OH)} \xrightarrow[\text{HCl}]{\text{ZnCl}_2}$$

EXPERIMENT

28   Aldehydes and Ketones

APPARATUS

Test tubes, test tube rack, 10 mL graduated cylinder, watch glass, Büchner funnel, filter paper, suction flask, heavy-walled rubber tubing, melting-point capillaries.

REAGENTS

Acetone, 95% ethanol, benzaldehyde, butanone, 3-pentanone, cyclohexanone, 2-propanol, iodine-potassium iodide solution, 3 M NaOH solution, 0.3 M AgNO$_3$ solution, 6 M NH$_3$ solution, 2,4-dinitrophenylhydrazine reagent, grease marking pencil (four per class), vials containing compounds listed in Table 28-1 for use as unknowns.

INTRODUCTION

A carbonyl group, which consists of a carbon-oxygen double bond, is extremely important in organic chemistry. This group may well be the most prevalent functional group in naturally occurring compounds. Various carbonyl compounds are possible, because several different atoms may bond to the carbon atom of the carbonyl group.

| Carbonyl group | Aldehyde | Ketone |

When one hydrogen atom is bonded to the carbon atom, the compound is an aldehyde. The other group can be a hydrogen atom, an alkyl group (R), or an aromatic group (Ar). Ketones are compounds in which the carbonyl group is bonded to two other carbon-containing groups. These bonded groups may be alkyl or aromatic groups.

The carbonyl group of both aldehydes and ketones react with primary amines to form imines.

an imine

Aldehydes and Ketones

When the amine is 2,4-dinitrophenylhydrazine, the product is a bright yellow to orange solid. This reaction is useful in identifying unknown aldehydes and ketones, because 2,4-dinitrophenyl-hydrazones (or 2,4-DNP products) have characteristic melting points.

2,4-Dinitrophenylhydrazine   A 2,4-Dinitrophenylhydrazone

Aldehydes and ketones may be distinguished based on the difference in their behavior toward oxidizing agents. Aldehydes are easily oxidized at the hydrogen atom attached to the carbonyl group. Ketones are not easily oxidized, since the reaction would require breaking a carbon-carbon bond. Mild oxidizing agents will then oxidize aldehydes but not ketones.

Tollens' reagent is a mild oxidizing agent that is an alkaline solution of $Ag(NH_3)^+$. When an aldehyde is oxidized by Tollens' reagent, metallic silver is deposited as a mirror on the walls of a test tube. The silver mirror is formed only if the glass is very clean. Otherwise, the silver is formed as a dark precipitate. The organic product is a carboxylate ion. The unbalanced equation is

$$R-\overset{\overset{\displaystyle O}{\|}}{C}-H \ + \ Ag(NH_3)_2^+ \longrightarrow R-\overset{\overset{\displaystyle O}{\|}}{C}-O^- \ + \ Ag$$

The presence of a methyl group directly bonded to the carbonyl group can be determined by treatment with a solution of I_2 and NaOH. The three hydrogen atoms of the methyl group are replaced by iodine. The resulting triiodomethyl group is cleaved by the base to produce a carboxylate ion and iodoform.

$$R-\overset{\overset{\displaystyle O}{\|}}{C}-CH_3 \ + \ 3 \ I_2 \ + \ 3 \ OH^- \longrightarrow R-\overset{\overset{\displaystyle O}{\|}}{C}-CI_3 \ + \ 3 \ I^- \ + \ 3 \ H_2O$$

$$R-\overset{\overset{\displaystyle O}{\|}}{C}-CI_3 \ + \ OH^- \longrightarrow R-\overset{\overset{\displaystyle O}{\|}}{C}-O^- \ + \ CHI_3$$

Iodoform, which is formed as a yellow precipitate, melts at 119°C. The formation of the yellow solid is visual proof of the presence of a methyl ketone in the structure.

$$R-\overset{\overset{\displaystyle O}{\|}}{C}-CH_3$$

a methyl ketone

Under the conditions of the iodoform reaction, alcohols with a methyl group attached to the carbon containing the –OH group also produce iodoform. This result is obtained because such alcohols are oxidized by the reagent to produce a carbonyl compound.

218

OH
|
—CH–CH₃ ⟶ —C̈–CH₃ $\xrightarrow[\text{NaOH}]{I_2}$ —C̈–O⁻ + CHI₃

PROCEDURE

Study Section K, "Melting Point," on pages 22-24 prior to starting this experiment.

CAUTION: The 2,4-dinitrophenylhydrazine test solution will stain your skin yellow. The Tollens' test solution will stain your skin black. Avoid spilling these chemicals on your skin.

Obtain from your laboratory instructor an unlabeled sample of an aldehyde or ketone. Your unknown will be one of the compounds listed in Table 28-1. Run tests to characterize the compound as an aldehyde or ketone and to characterize any other structural features of the unknown. Determine the melting point of the 2,4-DNP product of the unknown.

Table 28-1. Properties of selected aldehydes and ketones

Name	Formula	Boiling point (°C)	Melting point of 2,4-dinitrophenylhydrazone (°C)
Acetaldehyde (ethanal)	CH_3CHO	21	147
Hexanal	$C_5H_{11}CHO$	128	104
Heptanal	$C_6H_{13}CHO$	156	109
Benzaldehyde	C_6H_5CHO	179	237
Acetone	CH_3COCH_3	56	126
Butanone	$CH_3COC_2H_5$	80	117
2-Pentanone	$CH_3COC_3H_7$	102	144
3-Pentanone	$C_2H_5COC_2H_5$	102	156
4-Methyl-2-pentanone	$CH_3COC_4H_9$	119	95
4-Heptanone	$C_3H_7COC_3H_7$	151	75
Cyclohexanone	$C_5H_{10}CO$	155	162

Part A—Dinitrophenylhydrazine Test

Dissolve several drops of carbonyl compound in 4 mL of 95% ethanol, and add about 2 mL of the 2,4-dinitrophenylhydrazine reagent. Allow the mixture to stand at room temperature. The 2,4-DNP product should crystallize within 5-10 minutes. Carry out this reaction with acetone, butanone, cyclohexanone, benzaldehyde, and with your unknown. The melting point of the 2,4-DNP product should help you to identify the unknown. Collect the product using vacuum filtration. While the 2,4-DNP product is drying, perform the Tollens' test and the iodoform reaction as described in Parts B and C. Then determine the melting point of the dry 2,4-DNP product.

Part B—Iodoform Reaction

Perform this test with acetone, butanone, 3-pentanone, cyclohexanone, 95% ethanol, 2-propanol, and your unknown. Dissolve 4 drops of the compound to be tested in 1 mL of methanol. Add 1 mL of water and 2 mL of 3 *M* NaOH solution. Add I$_2$–KI solution dropwise until a definite yellow color persists for at least 2 minutes after mixing. Record your observations after 5 minutes. The formation of a precipitate indicates either a methyl ketone or an alcohol that can be oxidized to a methyl ketone.

Part C—Tollens' Test

Prepare the Tollens' solution as follows. To a clean micro test tube add 2 mL of 0.3 *M* silver nitrate solution and 1 drop of 3 *M* sodium hydroxide solution. Note the formation of the precipitate (Ag$_2$O). Now add dropwise about 6 drops of 6 *M* aqueous ammonia until the precipitate just dissolves. Carefully stopper the test tube and shake it until the precipitate just dissolves. Add more 6 *M* aqueous ammonia if necessary until the precipitate just dissolves. Do not add an excess. Add 2 drops of the compound to be tested to the Tollens' solution, and allow the mixture to stand at room temperature for 10 minutes. The formation of a silver mirror on a portion of the wall of the test tube or of a large amount of finely divided, metallic silver black precipitate is a positive test for an aldehyde. Perform this test with acetone, 95% ethanol, benzaldehyde, and your unknown.

CAUTION: All ammoniacal solutions of silver should be placed in the proper receptacle for disposal. Prolonged standing of these salts produces explosive compounds.

Report Sheet

_____   _____   _____
Name   **Instructor/Section**   **Date**

Part A—2,4-Dinitrophenylhydrazine Test

Compound	Observations
Acetone	_____
Butanone	_____
Cyclohexanone	_____
Benzaldehyde	_____
Unknown	_____

Melting point of 2,4-dinitrophenylhydrazone derivative of the unknown _____

Part B—Iodoform Test

Compound	Observations
Acetone	_____
Butanone	_____
3-Pentanone	_____
Cyclohexanone	_____
Ethanol (95%)	_____
2-Propanol	_____
Unknown	_____

Part C—Tollens' Test

Compound	Observations
Acetone	_____
Ethanol (95%)	_____
Benzaldehyde	_____
Unknown	_____

Conclusions concerning the identity of the unknown _____

Identity of the unknown compound _____

Describe how your observations support your conclusion about the identity of the unknown.

Answer the questions on the next page.

QUESTIONS

1. Write the balanced equation for the reaction that occurs between 2,4-dinitrophenylhydrazine and cyclopentanone.

2. Write the balanced equation for the reaction that occurs between the Tollens' reagent and benzaldehyde.

3. Discuss the result you would obtain if you ran the iodoform test on 2-pentanol.

4. The boiling points of 2-pentanone and 3-pentanone are identical. Indicate how the preparation of the 2,4-dinitrophenylhydrazone derivative is useful in identifying the two ketones.

5. An unknown liquid does not react when heated with an ammoniacal silver nitrate solution. The compound gives a positive iodoform test. The 2,4-dinitrophenylhydrazone derivative of the same unknown had a melting point of 115-117°C. Identify the unknown. Discuss the reasons for your identification.

EXPERIMENT

29 Carboxylic Acids and Derivatives

APPARATUS

Boiling chips, reflux apparatus, 16 × 150 mm test tubes, suction flask, Büchner funnel, 25 mL Erlenmeyer flask, ring stand, iron ring, wire gauze, clamp, burner, rubber tubing, matches, filter paper.

REAGENTS

Acetic acid, benzoic acid, stearic acid, salicylic acid, methanol, absolute ethanol, acetyl chloride, ethyl benzoate, 3 M sodium hydroxide, concentrated sulfuric acid, solid sodium hydroxide, litmus paper.

INTRODUCTION

Carboxylic acids and their derivatives are compounds in which a carbonyl carbon atom is bonded to an atom or group of atoms other than carbon or hydrogen.

| Acid derivative | Carboxylic acid | Ester | Acid chloride |

Acids have a hydroxyl group, esters have an alkoxy group, and acid chlorides have a chloride bonded to the carbonyl carbon atom. Since these compounds are structurally related, they are interconvertible by appropriate substitution reactions.

Carboxylic acids form hydrogen bonds with water and are soluble providing the alkyl group is not too large. Carboxylic acids are weak acids and are only slightly ionized in water. However, reaction with sodium hydroxide converts carboxylic acids into carboxylate salts. The salts are more soluble than the carboxylic acids because they are ionic.

Carboxylic acids react with alcohols in the presence of a strong acid as a catalyst to form esters. The esterification reaction is favored by increased concentrations of one or both of the reactants.

$$R-\overset{\displaystyle O}{\underset{\displaystyle OH}{C}} \quad + \quad R'OH \quad \overset{H_3O^+}{\rightleftharpoons} \quad R-\overset{\displaystyle O}{\underset{\displaystyle O-R'}{C}} \quad + \quad H_2O$$

An ester can be cleaved into its component carboxylic acid and alcohol by the reversal of the reaction above in an excess of water. However, the reaction is made essentially quantitative by adding molar amounts of strong base. This reaction, called saponification, results in the formation of an alcohol and a carboxylate salt.

$$R-\overset{\displaystyle O}{\underset{\displaystyle OR'}{C}} \quad + \quad NaOH \quad \longrightarrow \quad R-\overset{\displaystyle O}{\underset{\displaystyle O^-}{C}} \quad + \quad Na^+ \quad + \quad R'OH$$

Acid chlorides are extremely reactive compounds that react to replace the chlorine atoms by substitution. Reaction with water results in the formation of an acid, whereas reaction with an alcohol results in the formation of an ester.

$$R-\overset{\displaystyle O}{\underset{\displaystyle Cl}{C}} \quad + \quad H_2O \quad \longrightarrow \quad R-\overset{\displaystyle O}{\underset{\displaystyle OH}{C}} \quad + \quad HCl$$

$$R-\overset{\displaystyle O}{\underset{\displaystyle Cl}{C}} \quad + \quad ROH \quad \longrightarrow \quad R-\overset{\displaystyle O}{\underset{\displaystyle OR}{C}} \quad + \quad HCl$$

PROCEDURE

Before starting this experiment study Sections C, "Testing for Odors (Wafting)" (pages 8-9); D, "Filtration" (pages 9-11); and F, "Heating Liquids" (pages 12-14).

The structures of the compounds used in this experiment are

$$CH_3-\overset{\displaystyle O}{\underset{\displaystyle OH}{C}}$$
Acetic acid

$$CH_3-\overset{\displaystyle O}{\underset{\displaystyle Cl}{C}}$$
Acetyl chloride

$$CH_3(CH_2)_{15}CH_2-\overset{\displaystyle O}{\underset{\displaystyle OH}{C}}$$
Stearic acid

Benzoic acid

Salicylic acid

Ethyl benzoate

Part A—Solubility of Carboxylic Acids

Determine the solubility of acetic acid, benzoic acid, and stearic acid, one at a time, in each of these solvents: ethanol, water, and 3 M sodium hydroxide solution. To determine the solubility of acetic acid, add 6 drops to 3 mL of the solvent in a 16 × 150 mm test tube. Shake the tube. If the acid dissolves, continue dropwise addition until no more dissolves or until the total volume reaches 4 mL. Classify the compound as soluble, slightly soluble, or insoluble. Record your observations on your Report Sheet. Repeat the solubility tests with benzoic acid and stearic acid, but use one-half a micro spatula of the compound. Dissolve approximately 1 g of benzoic acid in 50 mL of water by heating the solution to about 80°C.

CAUTION: Ethanol is flammable. Make sure that your ethanol solvent used earlier is in a stoppered container and is some distance from the burner.

Now set the solution aside to cool in cold water. Record your observations on your Report Sheet.

Part B—Esterification

CAUTION: Sulfuric acid is a strong corrosive acid that will cause severe skin burns. Immediately wash any affected areas with large quantities of water. Notify your instructor and seek a medical evaluation of the chemical burn.

CAUTION: Note the odor by wafting the vapors toward your nose.

1. Pour 1 mL of acetic acid and 1 mL of ethanol into a 16 × 150 mm test tube. Add about 4 drops of concentrated sulfuric acid very carefully, and mix. Stand the test tube in a beaker of hot water at about 80°C for 2 to 3 minutes. Record your observations on the Report Sheet. Write the equation for the reaction.

2. Gently warm a 16 × 150 mm test tube containing 1 g of salicylic acid, 2 mL of methanol, and 4 drops of concentrated sulfuric acid. After a few minutes, allow the mixture to cool and note the odor by wafting the vapors toward your nose. Record your observations on your Report Sheet. The odor is sometimes more apparent if the liquid is poured into about 25 mL of cold water. Write the equation for the formation of the product, oil of wintergreen.

Part C—Acid Chlorides

CAUTION: Acetyl chloride is an eye and skin irritant. Care should be exercised in handling it.

In the hood, carefully add acetyl chloride dropwise to 5 mL of water, and note the temperature of the test tube as the reaction proceeds. Test the aqueous solution with blue litmus paper. Record your observations on your Report Sheet, and write an equation for the reaction that occurred.

Part D—Saponification of an Ester

CAUTION: Sodium hydroxide is a caustic substance. Do not touch this substance. Rinse your skin with water immediately if it has a soapy feeling. Sodium hydroxide will cause irreversible eye damage and blindness. Ask your instructor to clean up any spilled material at once.

In a 125 mL Erlenmeyer flask, mix 5 mL of ethyl benzoate with a solution of 3 g of sodium hydroxide in 15 mL of water. Fit the flask with a reflux condenser as shown in Figure 14 on page 14. A few boiling chips are placed in the flask to prevent bumping. Ask your instructor to check your setup prior to lighting the burner. Boil the mixture gently for at least 15 minutes or until the ester layer has disappeared.

Cool the contents of the flask, dilute with about 50 mL of water, and add dilute hydrochloric acid until the solution is acidic to litmus paper. Filter the precipitate and recrystallize it from water by dissolving it in a minimum amount of hot water and allowing the solution to cool. Collect the solid by suction filtration, and compare its appearance with that of the benzoic acid from Part A. Record your observations on your Report Sheet. What is the other organic compound obtained from this reaction? Write an equation for the saponification of ethyl benzoate.

Report Sheet

Name   Instructor/Section   Date

Part A—Solubility of Carboxylic Acids

Compound	In ethanol	In water	In 3 M NaOH
Acetic acid			
Benzoic acid			
Stearic acid			

How is the solubility of benzoic acid in water affected by temperature?

Describe the crystals of benzoic acid.

Part B—Esterification

Reaction	Odor	Equation
Ethanol + acetic acid		
Methanol + salicylic acid		

Part C—Carboxylic Acid Derivatives

Reactants	Odor	Litmus test	Products
Acetyl chloride and water			

What do you conclude about the reactivity of acetyl chloride with water?

Part D—Saponification of an Ester

Reaction	Observations	Equation
Ethyl benzoate and sodium hydroxide		

Answer the questions on the next page.

QUESTIONS

1. What are the products of the reaction in Part D prior to the addition of dilute hydrochloric acid?

2. How could you quantitatively separate *p*-ethylbenzoic acid from ethyl benzoate and recover each material in pure form?

3. Write the products expected from the following reactions.

a. 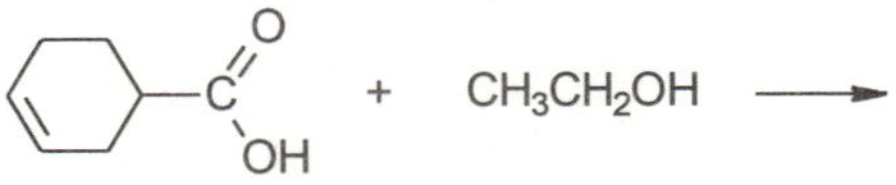 + CH₃CH₂OH ⟶

b. 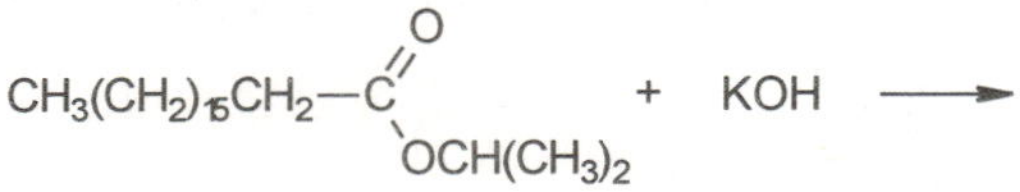 CH₃(CH₂)₆CH₂—C(=O)OCH(CH₃)₂ + KOH ⟶

c. + CH₃CH₂CH₂OH ⟶

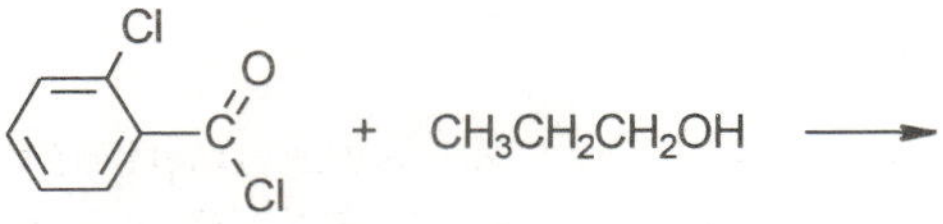

d. 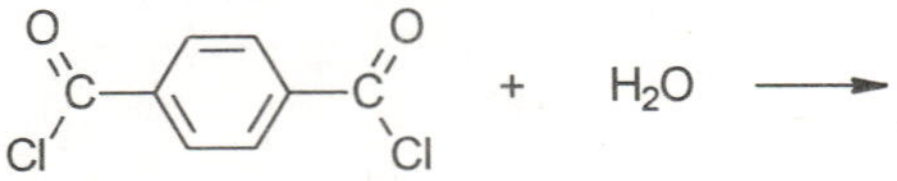 + H₂O ⟶

e. 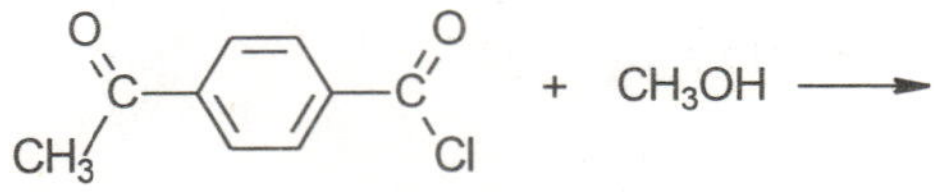 + CH₃OH ⟶

EXPERIMENT

30   Synthesis of *p*-Acetanisidine

APPARATUS

Beakers, ring stand, iron ring, clamp, wire gauze, 10 mL graduated cylinder, Büchner funnel, filter flask, rubber tubing, burner, matches, thermometer, melting-point apparatus, melting point tubes, filter paper (Whatman No. 1 or equivalent), plastic bags, labels. .

REAGENTS

Activated charcoal, p-anisidine, solid sodium acetate, acetic anhydride, concentrated hydrochloric acid, ice.

INTRODUCTION

The preparation of *p*-acetanisidine involves treating an amine with acetic anhydride to form an amide. For this experiment, the amine *p*-anisidine is treated with acetic anhydride forming p-acetanisidine.

1.34

123.15   *165.19*   *MP = 130 - 132*

p-Acetanisidine is similar in structure to three drugs which function as analgesics (pain relievers). These are acetanilide, acetaminophen, and phenacetin.

Acetanilide   Acetaminophen   Phenacetin

Synthesis of p-Acetanisidine

Note that these four compounds are amides that differ only in the functional group located at the para position relative to the amide. The simplest compound, which has a hydrogen atom at this position, is acetanilide. It was one of the first amides proven to relieve pain. However, serious side effects made it an undesirable medication. The para-hydroxyl derivative of acetanilide, acetaminophen, is also an effective antipyretic and analgesic. It is so popular that it has over fifty trade names. One common name is "Tylenol." p-Acetanisidine, which has a methoxy group at the para position, has little benefit as an analgesic. Phenacetin, which has an ethoxy group at the para position, has been reported to be a carcinogen. Prior to this report, phenacetin had been widely used to relieve pain. Phenacetin was a component of APC tablets. The initials APC indicate the three components in the tablet: aspirin, phenacetin, and caffeine.

Comparison of the effect of the substituted acetanilides when administered as a drug demonstrates a standard technique for improving drugs. A known analgesic such as acetanilide is modified by varying the structure. Each modified drug is then tested for effectiveness and for the absence of side effects. Each of the acetanilides can be prepared in the same way by reacting acetic anhydride with the appropriate amine.

You will be asked to calculate the percent yield for the p-acetanisidine produced and isolated. The reagent acetic anhydride is added in an amount more than sufficient to react with all of the p-anisidine. Thus, the p-anisidine is the limiting reagent. The maximum mass in grams of p-acetanisidine that can be made from the weighed amount of the limiting reagent p-anisidine is the theoretical yield. The theoretical yield can be calculated from the mass of p-anisidine and the chemical equation for the reaction.

The actual yield is the mass in grams of product isolated from the reaction mixture. The percent yield is calculated from the following formula.

$$\frac{\text{Actual yield}}{\text{Theoretical yield}} \times 100\% = \text{percent yield}$$

For many reactions it is not possible, even for the most experienced chemist, to attain a high percent yield. These reactions are incomplete or are accompanied by other reactions that use up the limiting reagent without producing the desired product. You should be careful in carrying out this synthesis so that you can isolate and turn in to your instructor a good yield of pure p-acetanisidine.

PROCEDURE

Study the following sections in this manual before starting this experiment: K, "Melting Point" (pages 22-24), and D, "Filtration" (pages 9-11).

CAUTION: Concentrated HCl and p-anisidine are corrosive chemicals. If they are spilled on your skin, flush the affected areas with water. Then inform your instructor to obtain further directions.

To a 100 mL beaker add 25 mL of water, then 40 drops of concentrated hydrochloric acid. Carefully weigh to the nearest one-hundredth of a gram 2.0-2.5 g of p-anisidine, then add it to the beaker and stir until the solid dissolves. If the solution of p-anisidine is discolored, add 0.3 g of activated charcoal, stir the mixture for a few minutes, and filter the mixture using gravity filtration. Add another 0.1 g of activated charcoal to the filtrate if it is still discolored and filter the mixture again.

need charcoal - should be clear soln

- not purple

230

Dissolve 3.2 g of sodium acetate crystals ($CH_3CO_2Na \cdot 3H_2O$) in 10 mL of water. Filter the sodium acetate solution if necessary to remove undissolved particles.

Pour the solution of p-anisidine hydrochloride into a 250 mL beaker supported on a wire gauze, and gently warm the solution over a low flame while stirring it with a glass rod. When the temperature of the solution reaches 50°C, remove the beaker from the ring, place it on the benchtop, and add 2.4 mL of acetic anhydride (1.08 g/mL).

CAUTION: The vapors of acetic anhydride are very irritating to the nose, throat, and eyes. Do this part of the experiment in the hood.

Stir the solution until the acetic anhydride dissolves and <u>immediately</u> add 10 mL of the sodium acetate solution. Stir the mixture vigorously to start the precipitation of the p-acetanisidine.

Cool the reaction mixture in an ice bath, and continue to stir the mixture while the product crystallizes. Collect the crystals by filtering them with suction on the Büchner funnel, wash them with 10 mL of cold water, and allow them to dry for a week..

Weigh the dried product, and determine its melting point. Transfer your product from the paper into a clean, dry, labeled, weighed (tared) plastic bag and turn it in with your written report. Your sample should be labeled as follows.

Date   Tare wt.
Sample wt.
Chemical name
Chemical formula
Melting point range
Your name

Report Sheet _____ _____ _____

Mass of *p*-anisidine   (1) _____

Volume of acetic anhydride   (2) _____

Mass of acetic anhydride   (3) _____

Mass of *p*-acetanisidine   (4) _____

Melting point of *p*-acetanisidine   (5) _____

Theoretical yield of *p*-acetanisidine   (6) _____

% yield of *p*-acetanisidine   (7) _____

Show your calculations. Answer the questions on the next page.

Synthesis of p-Acetanisidine

QUESTIONS

1. How would the percent yield be affected if the sample had not dried completely?

2. How would the percent yield be affected if an excess of acetic anhydride were not used and the reaction were incomplete?

3. What starting reactants would be needed to prepare the following compound?

4. What functional groups are present in a molecule of acetaminophen?

~~5.~~ Write the equation for the reaction of p-anisidine with hydrochloric acid.

~~6.~~ Would a solution of acetaminophen be acidic, neutral, or basic? Why?

5. How does your MP of acetanisidine compar to the known value (MP = 130-132). What conclusions might you make?

EXPERIMENT
31
Amino Acids and Proteins

APPARATUS

10 × 75 mm test tubes, 16 × 150 mm test tubes, medicine dropper, reflux condenser, iron ring, wire gauze, ring stand, clamp, Büchner funnel, suction flask, funnel, filter paper, rubber tubing, buret clamps, 125 mL Erlenmeyer flask, burner, matches, pH paper or litmus paper, boiling stones.

REAGENTS

Glycine, L-lysine, L-tyrosine, L-glutamic acid, casein (milk protein), urea, egg albumin, egg-white solution, 3 M sodium hydroxide, concentrated nitric acid, concentrated hydrochloric acid, 3 M hydrochloric acid, 6 M hydrochloric acid, 0. 1 M copper(II) sulfate, 0. 1 M manganese(II) nitrate, 0.1 M iron(II) chloride, 0.1 M lead(II) nitrate, 0.1 M silver nitrate, 5% sodium nitrite, activated charcoal, lead acetate paper.

INTRODUCTION

Amino acids contain both the carboxyl group and the amino group. Most amino acids of interest in biochemistry are α-amino acids; that is, the amino group is attached to the α-carbon atom. The chemical properties of the amino acids reflect both functional groups. Thus, they may act as either bases or acids.

$$R-\underset{\underset{NH_3^+}{|}}{C}HCO_2H \underset{H_3O^+}{\overset{OH^-}{\rightleftharpoons}} R-\underset{\underset{NH_2}{|}}{C}HCO_2H \underset{H_3O^+}{\overset{OH^-}{\rightleftharpoons}} R-\underset{\underset{NH_2}{|}}{C}HCO_2^-$$

Solutions of many amino acids are nearly neutral, since they contain both basic and acidic functional groups. If the alkyl group R– contains a second amino or carboxyl group, an aqueous solution of the amino acid will be alkaline or acidic, respectively. In this experiment you will examine the acid-base properties of amino acids.

Primary amines and amino acids containing a free –NH_2 group react with nitrous acid to evolve nitrogen gas. Secondary or tertiary amines do not yield nitrogen. This reaction will be studied in a qualitative manner in this experiment.

$$R-\underset{\underset{NH_2}{|}}{C}HCO_2H + HNO_2 \longrightarrow R-\underset{\underset{OH}{|}}{C}HCO_2H + N_2 + H_2O$$

Any additional structural features contained in the alkyl side chain behave in their own characteristic manner. Two naturally occurring amino acids, cysteine and cystine, contain sulfur. The presence of sulfur in compounds may be confirmed by generating H_2S from them. Hydrogen sulfide can be detected by the use of paper impregnated with lead acetate, $Pb(C_2H_3O_2)_2$. Lead(II) forms a dark PbS compound by reacting with H_2S.

Aromatic rings in amino acids react with electrophilic reagents, as do all aromatic compounds. For example, the nitration of aromatic compounds yields yellow nitrated products, which often turn deeper yellow when treated with a base. This procedure is called the xanthoproteic test.

A protein such as casein is a natural polymer that yields a mixture of amino acids upon hydrolysis. The properties of the hydrolysis mixture are simply those of the constituent amino acids.

A protein molecule

Biuret is obtained by heating urea.

Biuret contains multiple amide linkages that are similar to those contained in proteins. Biuret reacts with copper(II) ions in basic solution to form a reddish violet complex ion. Structures that resemble biuret, such as those partial structures listed below, also react with copper(II) to form colored complex ions.

Structures of Some Amino Acids

H—CHCO$_2$H
|
NH$_2$
Glycine

CH$_3$—CHCO$_2$H
|
NH$_2$
Alanine

HS—CH$_2$—CHCO$_2$H
|
NH$_2$
Cysteine

NH$_2$CH$_2$CH$_2$CH$_2$CH$_2$—CHCO$_2$H
|
NH$_2$
Lysine

HO$_2$CCH$_2$CH$_2$—CHCO$_2$H
|
NH$_2$
Glutamic acid

HO—⟨benzene ring⟩—CH$_2$CHCO$_2$H
|
NH$_2$
Tyrosine

PROCEDURE

Part A—Solubility and Acidity

1. Add a few crystals of glycine to 2 mL of distilled water in a micro test tube. Test the acidity of the solution with pH paper. Carry out the same experiments with a few crystals of L-lysine, L-tyrosine, and L-glutamic acid. Record your observations on the Report Sheet, and indicate the reason for the observed behavior.
2. Prepare a suspension of 0.1 g of L-tyrosine in 2 mL of water in a 16 × 150 mm test tube. Add 1 mL of 3 M sodium hydroxide, and shake the test tube. Add a small piece of litmus paper to the solution, and then add 3 M hydrochloric acid dropwise until the solution is slightly acidic. Shake the tube for a minute, and record your observations. Add 10 more drops of acid, shake the tube, and note any changes that occur. Record your observations on the Report Sheet.
3. Prepare a casein solution from 0.1 g of casein, 5 mL of water, and 2 mL of 3 M sodium hydroxide. Stopper the test tube, and shake it until a colloidal suspension is obtained. Save half the suspension for Part C. To the other half, add 3 mL of concentrated hydrochloric acid dropwise. Observe the result. Stopper the tube and shake the tube and mixture well. Save this solution for Part B.

CAUTION: Hydrochloric acid is a strong acid that will cause severe skin burns. Immediately wash any affected areas with large quantities of water. Notify your instructor and seek a medical evaluation of the chemical burn.

Part B—Reaction with Nitrous Acid

1. In one 16 × 150 mm test tube place 5 mL of 3 M hydrochloric acid, and in another tube put 0.1 g of glycine with 5 mL of 3 M hydrochloric acid. Allow both tubes to cool to 0°C in an ice bath. Then carefully add 1 mL of 5% sodium nitrite to each test tube. Record your observations on the Report Sheet.
2. Repeat the above test, using the casein solution in hydrochloric acid prepared in Part A(3). Remember to cool the contents of the test tube before adding sodium nitrite. Record your observations on the Report Sheet.

Part C—Biuret Test

1. Heat a dry test tube containing 0.5 g of urea until the solid melts and a gas is evolved. Cautiously note the odor of the gas, and test it with a piece of moist litmus paper held near the mouth of the test tube. Heat the contents gently until gas evolution ceases. Cool the tube and dissolve the resultant solid in 3 mL of hot distilled water. Add 2 mL of 3 M sodium hydroxide and 2 drops of 0.1 M copper sulfate. Shake the tube, and observe any changes. Record your observations on the Report Sheet.
2. Dissolve 0.5 g of urea in 2 mL of water, and add 2 mL of 3 M sodium hydroxide and 2 drops of 0.1 M copper sulfate. Compare the results with those of the previous experiment. Record your observations on the Report Sheet.
3. To the casein solution in base prepared in Part A(3) add 2 mL of water and 2 drops of 0.1 M copper sulfate solution. Shake the solution, and observe the results. Record the results on the Report Sheet.

Part D—Hydrolysis of a Protein

1. Read Section F, "Heating Liquids," on pages 12-14 to review how to set up a reflux condenser, and study Figure 14. Construct a reflux apparatus using a 125 mL Erlenmeyer flask and a small condenser. Add 25 mL of 6 M hydrochloric acid and 0.5 g of casein. A porous chips or boiling stones added to the flask will control boiling. Before continuing this experiment, have your instructor check your apparatus. Gently heat the mixture under reflux for about 30 minutes using a low burner flame. Allow the solution to cool to room temperature. Add about 0.3 g of activated charcoal, and swirl the mixture for 2 minutes. Using suction, filter the solution, and save the filtrate that contains hydrolyzed casein in a stoppered Erlenmeyer flask.

    Perform the nitrous acid test as follows. Cool a 5 mL portion of the hydrolyzed casein in an ice-water bath, and add 1 mL of 5% sodium nitrite. Record your observations on the Report Sheet.

    Perform the biuret test as follows. Carefully neutralize a 2 mL portion of the hydrolyzed casein with 3 M sodium hydroxide and then add an additional 3 mL. Add 2 drops of 0.1 M copper sulfate solution. Record your observations on the Report Sheet and compare the results of both the biuret test and the nitrous acid test with those obtained for casein from Parts B and C.

Part E—Xanthoproteic Test

CAUTION: Nitric acid is a strong corrosive acid that will cause severe skin burns. Immediately wash any affected areas with large quantities of water. Notify your instructor and seek a medical evaluation of the chemical burn.

1. To 2 mL of egg-white solution in a test tube, add 2 mL of concentrated nitric acid. Warm the test tube gently, and note any changes. Cool the solution and cautiously add 3 M sodium hydroxide until slightly basic. Record your observations on the Report Sheet.
2. Repeat the above procedure with 0.1 g of casein. Record your observations on the Report Sheet.

Part F—Test for Sulfur

To 0.2 g of dried egg albumin in a 25 × 200 mm test tube, add 10 mL of 3 M sodium hydroxide. Gently boil the solution for 15 minutes, but remove the heat if foaming threatens to discharge the contents. Cool the solution in an ice bath until it reaches room temperature. Acidify the solution with 3 M hydrochloric acid. Again boil the contents after placing a piece of moist lead acetate paper over the mouth of the test tube. Record your observations on the Report Sheet.

Part G—Precipitation with Metal Ions

Place 2 mL of egg-white solution in each of four test tubes. Slowly add 2 mL of 0.1 M silver nitrate solution to the first test tube and observe the result. Repeat the experiment with 0.1 M solutions of iron(III) chloride, copper(II) sulfate, and manganese(II) nitrate.

Report Sheet

Part A—Solubility and Acidity of Amino Acids

Amino acid	Solubility	pH	Remarks
Glycine			
Lysine			
Tyrosine			
Glutamic acid			

Describe the solubility behavior of tyrosine observed in Part A, paragraph 2.

Describe the solubility behavior of casein observed in Part A, paragraph 3.

Part B—Nitrous Acid Test

Substance	Observations
Hydrochloric acid	
Glycine	
Casein	

Part C—Biuret Test

Substance	Observations
Heated urea	
Unheated urea	
Casein	

Part D—Comparison of Casein with Hydrolyzed Casein

(For casein, copy your results from Parts B and C.)

	Observations	
Test	Casein	Hydrolyzed Casein
Nitrous acid	_____	_____
Biuret	_____	_____

Part E—Xanthoproteic Test

Substance	Observations
Egg white	_____
Casein	_____

Part F—Test for Sulfur

Description of gas evolved.

Equation for the reaction of the gas with lead acetate

Part G—Precipitation with Metal Salts

Metal salt	Observations
$AgNO_3$	_____
$FeCl_3$	_____
$CuSO_4$	_____
$Mn(NO_3)_2$	_____

Attach answers to the questions on the following pages.

QUESTIONS

1. Explain the observed pH of the solutions of the amino acids glutamic acid and lysine.

2. Write equations illustrating the equilibria that occur as a basic solution of tyrosine is reacted with HCl.

3. Explain the differences observed in the reactions of glycine and casein with nitrous acid.

4. Explain the differences observed in the reactions of casein and hydrolyzed casein with nitrous acid.

5. Explain the results of the biuret test with casein and hydrolyzed casein.

6. What gas was evolved from the acidified solution of hydrolyzed albumin? Write the equation for the reaction of the gas with lead(II) acetate.

7. Why is egg white suggested as an antidote for lead poisoning?

EXPERIMENT

32   Carbohydrates

APPARATUS

Fourteen 16 × 150 mm test tubes, 600 mL beaker, boiling chips, grease marking pencil, test tube rack, burner, iron ring, ring stand, clamp, rubber tubing, matches, wire gauze, 10 mL graduated cylinder, medicine dropper.

REAGENTS

1% stock solutions of solubilized starch ("Vitex"), glucose, fructose, maltose, sucrose, and xylose; Molisch reagent, concentrated sulfuric acid, 3 M hydrochloric acid, 3 M sodium hydroxide, iodine-potassium iodide solution, Benedict's reagent, Barfoed's reagent, Bial's reagent, Seliwanoff's reagent, cornstarch, unknown solid carbohydrate samples.

INTRODUCTION

Structurally, carbohydrates are polyhydroxyaldehydes, polyhydroxyketones, or compounds that are chemically convertible to these two components. Therefore, the chemistry of carbohydrates resembles chemical reactions of alcohols and aldehydes or ketones. Aldehyde containing carbohydrates are called aldoses, whereas ketonic carbohydrates are called ketoses. Carbohydrates containing five and six carbon atoms are called pentoses and hexoses, respectively.

The carbohydrates are divided into classes: monosaccharides, oligosaccharides (Greek oligos, a few), and polysaccharides. Monosaccharides contain a single carbohydrate unit; that is, they do not yield simpler carbohydrates upon hydrolysis. The oligosaccharides are subclassified as disaccharides, trisaccharides, and so forth, according to the number of monosaccharide units released upon hydrolysis. Polysaccharides contain hundreds to thousands of monosaccharide units.

Monosaccharides exist as cyclic hemiacetals derived from the intramolecular reaction of a hydroxyl and a carbonyl group. The major glucose isomer is β-D-glucopyranose; the minor isomer is α-D-glucopyranose. The structures of the α and β forms differ only at the configuration of the anomeric carbon atom, and these two molecules are called anomers. The anomeric carbon atom is the new chiral center generated by the hemiacetal formation. These two hemiacetals exist in equilibrium with each other in solution through a small concentration of the open-chain form. This equilibrium explains many of the chemical and physical properties of carbohydrates. The chemical reactions of glucose and other carbohydrates that parallel those observed for the aldehyde group arise from a small equilibrium concentration of the open-chain aldehyde form.

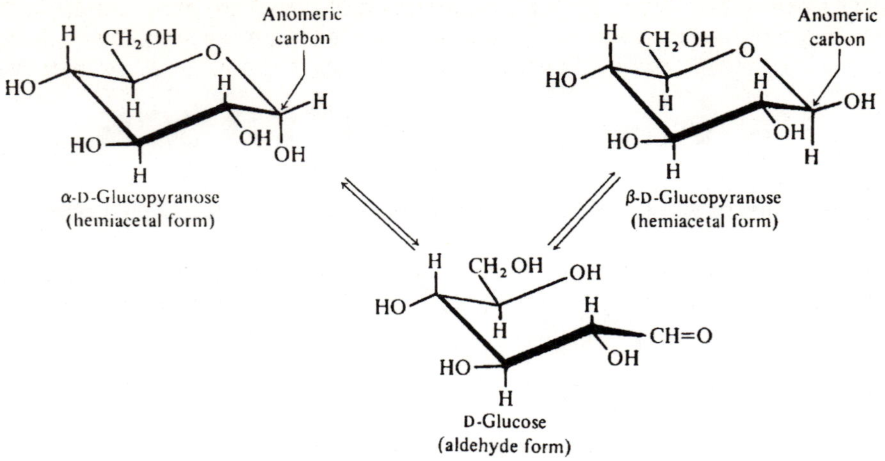

α-D-Glucopyranose
(hemiacetal form)

β-D-Glucopyranose
(hemiacetal form)

D-Glucose
(aldehyde form)

Disaccharides are compounds in which one monosaccharide, acting as an alcohol, reacts with the hemiacetal of the second monosaccharide to form an acetal. Maltose is a disaccharide containing two glucose units. The α-isomer of maltose is shown below. Equilibration of the α-maltose isomer through an open-chain form would yield the β-isomer.

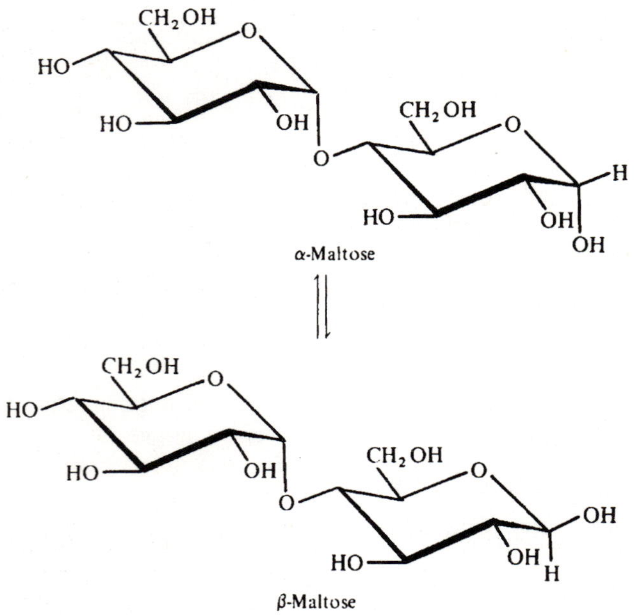

α-Maltose

β-Maltose

Disaccharides in which one monosaccharide unit contains a hemiacetal center will undergo oxidation. However, if the two monosaccharides are joined through the oxygen atom at their hemiacetal (or hemiketal) center to form an acetal or ketal, the disaccharide will not be oxidized. Thus, a useful method of determining structures of di- or polysaccharides is to treat them with mild oxidizing agents.

Open-chain aldehydes and α-hydroxyketones that are in equilibrium with the closed form readily undergo oxidation. Mild oxidizing agents such as Tollens' reagent (ammoniacal solution of silver ion) and Benedict's reagent (a copper citrate complex ion) are often used. Carbohydrates that give positive tests are termed reducing sugars. Note that, in contrast to simple ketones, α-hydroxyketones are readily oxidized because they can isomerize to α-hydroxyaldehyes.

Ketoses are α-hydroxyketones and thus show reactions that are ordinarily characteristic only of aldehyde groups, as in aldoses. Both aldoses and ketoses are oxidized by Tollens' reagent and Benedict's solution.

$$R\!-\!CHO + 2\,Ag(NH_3)_2^+ + 3\,OH^- \longrightarrow RCO_2^- + 4\,NH_3 + 2\,Ag + 2\,H_2O$$

$$R\!-\!CHO + 2\,Cu^{2+} + 5\,OH^- \longrightarrow RCO_2^- + Cu_2O + 3\,H_2O$$

Polysaccharides are high molecular mass polymers of monosaccharide units. Two of the most common polysaccharides are starch and cellulose.

Qualitative Tests for Carbohydrates

Copper Reduction Tests. The most common test for detecting reducing sugars involves heating the sample with an alkaline reagent containing cupric ions, Cu^{2+} as a citrate complex ion. The aldehyde group or α-hydroxyketone group is oxidized, and the copper is reduced to cuprous oxide (Cu_2O). In this experiment *Benedict's solution* is used to detect reducing sugars.
Barfoed's solution contains cupric ions in a slightly acidic medium. The milder conditions allow the differentiation between reducing monosaccharides and reducing disaccharides. If the time of heating is carefully controlled, disaccharides do not react while reducing monosaccharides give the positive result (red Cu_2O precipitate).

Furfural-Forming Tests. Carbohydrates undergo dehydration in the presence of nonoxidizing acids to form furfural or hydroxymethylfurfural.

Furfural   **Hydroxymethylfurfural**

Either of these aldehydes will condense with phenols to give intensely colored compounds. Under vigorous conditions (Molisch solution), all carbohydrates react. Under milder conditions (Bial's solution and Seliwanoff's solutions), only certain classes of carbohydrates react and these tests become more specific. The structures of the aromatic phenols used in a variety of reagents are:

α-naphthol   orcinol   resorcinol

Carbohydrates

The *Molisch* solution contains α-naphthol in ethanol. When combined with any carbohydrate and concentrated sulfuric acid, a purple condensation product is formed.

Bial solution contains orcinol in concentrated hydrochloric acid and a trace of ferric chloride. This solution reacts with pentoses to yield a green or blue condensation product. Hexoses yield a muddy brown to grey solution.

Seliwanoff solution contains resorcinol in 6 M hydrochloric acid. Ketoses (particularly those containing five or six carbon atoms) react to form cherry-red condensation products. Aldoses give a blue-green solution; they may give a peach color if not promptly observed.

Iodine-Potassium Iodide Test. A solution containing iodine and potassium iodide is added to a carbohydrate solution to test for starch. Formation of a deep blue color is a positive test.

PROCEDURE

Part A—Identification of an Unknown Carbohydrate

Obtain from your instructor an unknown sample. The order of tests that you should carry out and the results are given in Figure 32-1.

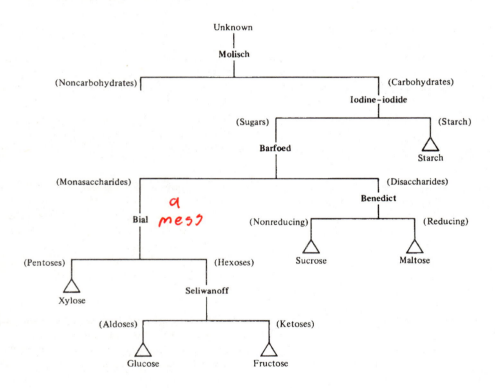

Figure 32-1. Flow chart for identification of unknown carbohydrate.

The solid will be either sucrose, maltose, xylose, glucose, fructose, or solubilized starch. You should be able to identify your unknown based on the results of these tests. It is advisable to run

246

each test on all seven solutions at once. After each test, compare the result for your unknown with the result for each known. The structural features of some carbohydrates are summarized in Table 32-1.

Table 32-1. Structural features of carbohydrates

Starch	polysaccharide	12 carbon atoms	many glucose units	nonreducing
Maltose	disaccharide	12 carbon atoms	two glucose units	reducing
Sucrose	disaccharide	12 carbon atoms	glucose and fructose units	nonreducing
Fructose	monosaccharide	6 carbon atoms	ketose	reducing
Glucose	monosaccharide	6 carbon atoms	aldose	reducing
Xylose	monosaccharide	5 carbon atoms	aldose	reducing

A 1% solution of the unknown is prepared by dissolving about 0.5 g in 50 mL of distilled water; 1% solutions of the known carbohydrates are provided in the laboratory. The basis for the identification is the result of each test for a specific structural feature. Each of the seven solutions (six known solutions plus your unknown) will be tested with Molisch reagent, iodine-potassium iodide reagent, Barfoed's reagent, Bial's reagent, Benedict's reagent, and Seliwanoff reagent.

Note that many tests require the use of boiling water. Prior to immersion into the boiling water, label each tube with a special marker you borrow from the instructor. Ordinary gummed labels will steam off and leave the tubes unidentified. The water will boil more smoothly when a boiling chip is present.

Molisch Test. The Molisch test is a diagnostic test for any carbohydrate. *pour slowly is key*

CAUTION: This test uses concentrated sulfuric acid (H$_2$SO$_4$) which can cause severe skin burns. Immediate washing of the affected area and prompt reporting to the instructor are important to avoid skin burns if the sulfuric acid is spilled.

Add 2 drops of Molisch reagent to 2 mL of each of the carbohydrate solutions in separate test tubes and mix well. Slowly pour each solution down the inside of another set of seven tubes containing 2 mL of concentrated sulfuric acid. Do not shake to mix the solutions. If the test tube containing the sulfuric acid is held at an angle of about 30°, two separate layers will form without appreciable mixing. Note the color change at the interface of the two liquids.

Iodine-Iodide Test. Add 2 drops of iodine-potassium iodide to 2 mL of each of the seven carbohydrate solutions. Note the color changes after mixing.

Barfoed's Test. Barfoed's reagent is useful in differentiating between reducing monosaccharides and reducing disaccharides. Place 3 mL of Barfoed's reagent and 1 mL of the carbohydrate solution in a test tube and place the tube in a beaker of boiling water for 5 minutes (no longer!). After heating, remove the test tube and cool it in running water. A dark red precipitate of cuprous oxide is a positive test for a reducing monosaccharide.

Bial's Test. This test can be used to distinguish between pentoses and hexoses.

CAUTION: This test uses concentrated (12 M) HCl which can cause severe skin burns. Immediate washing of the affected area and prompt reporting to the instructor are important to avoid skin burns if the hydrochloric acid is spilled.

I did not do - It going to be too messy.
I'll do it later

Carbohydrates

Place 3 mL of Bial's reagent and 2 drops of the 1% carbohydrate solution in a test tube and place the tube in a beaker of boiling water and heat. Remove each tube as soon as you see a color change. A pentose will provide a green or blue color within 2 minutes. The muddy brown or grey color develops from a ketose or disaccharide containing a ketohexose.

Benedict's Test. Benedict's test is used to identify reducing sugars. Add 10 drops of each of the seven 1% carbohydrate solutions to 3 mL of Benedict's solution contained in individual test tubes. Place all seven tubes in boiling water at the same time, and remove them after exactly 2 minutes. A dark red precipitate of cuprous oxide is a positive test.

Seliwanoff's Test. This test can be used to distinguish between aldohexoses and ketohexoses.

CAUTION: This test uses concentrated (6 M) HCl which can cause severe skin burns. Immediate washing of the affected area and prompt reporting to the instructor are important to avoid skin burns if the hydrochloric acid is spilled.

Place 10 drops of each carbohydrate solution, 15 drops of distilled water, and 9 mL of Seliwanoff's reagent into test tubes. Place in a beaker of boiling water, and heat no longer than 2.5 minutes. Remove the tubes from the hot water and note the colors. Ketohexoses produce a cherry-red product; aldoses give a blue-green solution.

Part B—Hydrolysis of Carbohydrates

The properties of hydrolyzed carbohydrates are compared with those of the unhydrolyzed parent material. For this part unsolubilized starch (corn starch) and sucrose are used.

Acid-Catalyzed Hydrolysis of Sucrose. Add 1 mL of 3 M hydrochloric acid to 10 mL of 1% sucrose solution. Place the mixture into a beaker of boiling water, and heat for 30 minutes. Cool the solution, and neutralize it to litmus with 3 M sodium hydroxide. Test the neutralized solution with Benedict's reagent. Run concurrently another Benedict's test on unhydrolyzed sucrose. Use the procedure for the Benedict's test in Part A.

Acid-Catalyzed Hydrolysis of Corn Starch. Mix 2 g of corn starch with 20 mL of cold water and crush the lumps with a stirring rod. Pour the suspension slowly into a beaker containing 250 mL of boiling distilled water. Boil the mixture for an additional 2 minutes, and allow it to cool. Hydrolyze 10 mL of the corn starch solution as follows. Add 1 mL of 3 M hydrochloric acid, and heat the mixture in a beaker of boiling water for 30 minutes. Cool the solution, and neutralize it to litmus with 3 M sodium hydroxide solution.

Perform the Benedict's test and the I_2–KI Test as described in Part A. Perform each test simultaneously on the unhydrolyzed cornstarch solution and on the hydrolyzed cornstarch solution. Compare the results.

Report Sheet

Name _____ _____ _____
Instructor/Section   Date

Part A—Identification of an Unknown Carbohydrate

	Molisch	Iodine-iodide	Barfoed	Bial	Benedict	Seliwanoff
Starch						
Maltose						
Sucrose						
Fructose						
Glucose						
Xylose						
Unknown						

Identification of unknown _____

Part B—Hydrolysis of Carbohydrates

State of Sucrose	Observations of Benedict's test
Unhydrolyzed sucrose	
Hydrolyzed sucrose	

State of Starch	Benedict's test	I_2 - KI test
Unhydrolyzed starch solution		
Hydrolyzed starch solution		

Answer the questions on the next page.

Carbohydrates

QUESTIONS

1. Name or describe a carbohydrate test that could be used to distinguish clearly between the following pairs.

    a. fructose and arabinose

    b. fructose and galactose

    c. maltose and mannose

    d. sucrose and lactose

    e. sucrose and starch

2. What structural features do all reducing sugars have in common?

3. Explain why sucrose reacts rapidly with Seliwanoff's reagent.

4. Explain the changes you observed in the I_2-KI test and the Benedict's test for corn starch and hydrolyzed corn starch.

5. A compound does not react with Barfoed's solution. It gives a brown color with the Bial solution and a cherry red color with Seliwanoff solution. Draw a structure for the compound.

Draw the reaction when Glucose is reacted with Benedict's Solution.

Fructose and Glucose are reacted Seliwanoff's Reagent. Compare. Predict the result How do your results differ from to the predicted results?

EXPERIMENT

33 Fats, Oils, Soaps, and Detergents

APPARATUS

16 × 150 mm test tubes, 10 × 75 mm test tubes, reflux apparatus, medicine dropper, glass stirring rod, beaker, boiling chips.

REAGENTS

Cottonseed oil, Crisco, hexane, 1,4-dioxane, 5% bromine solution in 1,4-dioxane, 95% ethanol, sodium hydroxide, concentrated sodium chloride solution, Ivory soap or flakes, Fab, 0.1 M calcium chloride, 0.1 M magnesium chloride, 0.1 M iron(III) chloride, 0.2 M ammonium molybdate solution, 6 M nitric acid, litmus paper or pH paper, mineral oil.

INTRODUCTION

Triacylglycerol, one of the many classes of lipids, are esters of the alcohol glycerol and long-chain carboxylic acids.

$$
\begin{array}{l}
CH_2-O-\overset{\displaystyle O}{\overset{\|}{C}}-R \\
| \quad\quad\quad O \\
CH-O-\overset{\displaystyle }{\overset{\|}{C}}-R' \\
| \quad\quad\quad O \\
CH_2-O-\overset{\displaystyle }{\overset{\|}{C}}-R''
\end{array}
$$

The triacylglycerols contain acids of twelve to eighteen carbon atoms and consist of mixtures of several different acids. If the acids are predominantly unsaturated, the triacylglycerol is a liquid and is classified as an oil. Triacylglycerols containing higher percentages of saturated acids have higher melting points and are known as fats.

Hydrolysis of a fat or oil with base to produce glycerol and salts of acids is known as saponification. Soaps are the salts of long-chain carboxylic acids. The sodium and potassium salts are soluble in water, whereas the magnesium, calcium, and iron salts are not.

Detergents are salts of aryl sulfonates or alkyl sulfates.

$$Ar—SO_3^- \, Na^+$$
Sodium aryl sulfonate

$$R—OSO_3^- \, Na^+$$
Sodium alkyl sulfate

Because the calcium salts of aryl sulfonates and alkyl sulfates are soluble in water, detergents can be used in hard water, whereas soaps would form insoluble precipitates with calcium ions.

Every year Americans use millions of kilograms of detergents to wash their clothes and dishes. Almost every gram of these products eventually finds its way into our lakes and streams. Many laundry products contain phosphates that, after entering the water system, stimulate the growth of algae. Phosphates act as a fertilizer for algae in a manner similar to the way they fertilize garden plants or grass. Algae consume large quantities of the oxygen dissolved in the water. The decreased concentration of dissolved oxygen suffocates fish and other animals living in the water and disturbs the ecological balance.

In this experiment, commercial laundry detergents are tested to determine if they contain phosphates. The basis for this test is the chemical reaction between the phosphate anion and the molybdate anion in acidic solutions.

$$12 \, Mo_7O_{24}^{6-} \; + \; 7 \, PO_4^{3-} \; + \; 72 \, H^+ \longrightarrow \; 7 \, PMo_{12}O_{40}^{3-} \; + \; 36 \, H_2O$$

The test solution of ammonium molybdate, $(NH_4)_6Mo_7O_{24}$, is added to a previously acidified sample. If the sample contains phosphate anions, a finely divided, bright yellow precipitate of ammonium phosphomolybdate, $(NH_4)_3PMo_{12}O_{40}$, is formed.

In this experiment the student is asked to bring to the laboratory a small sample of a powdered or liquid detergent. Be sure to check the label to determine whether or not the manufacturer admits putting phosphates in the product. You should also test a detergent from a classmate and discuss the results for other detergents with other class members.

PROCEDURE

Before starting this experiment, study Sections D, "Filtration" (pages 9-11), and F, "Heating Liquids" (pages 12-14).

Part A—Solubility of Cottonseed Oil

In each of four 16 × 150 mm test tubes place 0.5 mL (about 10 drops) of cottonseed oil. Add 1 mL of water to the first sample, ethanol to the second, hexane to the third, and 1,4-dioxane to the fourth. Shake each tube and its contents, and note the solubility in each solvent. Add an additional 5 mL of each solvent, and shake vigorously to determine if the oil will now dissolve. Record your observations on your Report Sheet.

Part B—Unsaturation of Triacylglycerols

Use 2 mL of the 1,4-dioxane solution of cottonseed oil prepared in Part A for this test of unsaturation. Add dropwise a solution of 5% bromine in 1,4-dioxane. Count the number of drops required to impart a bromine color to the solution. Dissolve 0.1 g of Crisco in 1 mL of 1,4-dioxane. Repeat the test for unsaturation. Record your observations on your Report Sheet.

Part C—Properties of Soaps and Detergents

CAUTION: Sodium hydroxide is caustic. Do not touch this substance. Rinse your skin with water immediately if it has a soapy feeling. Sodium hydroxide will cause irreversible eye damage and blindness. Clean up spilled particles at once.

Preparation. Dissolve 5 g of sodium hydroxide in 10 mL of distilled water and 20 mL of ethanol in a 125 mL Erlenmeyer flask. Add 5 g of Crisco and several boiling chips to the flask. Attach a reflux condenser as shown in Figure 14. Ask your instructor to check your apparatus. Reflux the contents of the flask for approximately 30 minutes or until the solution is clear and homogeneous. Dismantle the reflux apparatus. Remove the saponified mixture from the heat, cool the flask, and add about 80 mL of concentrated sodium chloride solution. Stir the mixture thoroughly and collect the precipitated soap by suction filtration.

Properties. Prepare a solution of 1 g of your prepared soap in 50 mL of boiling distilled water. Prepare similar solutions of a commercial soap and a detergent.

Alkaline Properties. Test each solution with litmus paper or pH paper. Record your observations on your Report Sheet.

Emulsifying Properties. Place 3 drops of mineral oil in each of four test tubes. Add 5 mL of distilled water to the first tube, 5 mL of your soap solution to the second, 5 mL of the commercial soap solution to the third, and 5 mL of detergent solution to the last tube. Shake each tube for a minute, and record your observations on your Report Sheet.

Effect of Metal Salts. Place 5 mL of your soap solution in each of three separate test tubes. Add 2 mL of a solution of 0.1 M calcium chloride to the first tube, 0.1 M magnesium chloride to the second, and 0.1 M iron(III) chloride to the third. Record your observations on your Report Sheet. Repeat these experiments with a commercial soap and a detergent.

Part D—Qualitative Test for Phosphate in a Detergent

The test for phosphate is very sensitive. Even a small amount of detergent remaining in a test tube after rinsing may give a positive test. Therefore, thoroughly rinse your tubes in distilled water or obtain new micro test tubes from the storeroom.

CAUTION: Nitric acid is a strong corrosive acid that will cause severe skin burns. Immediately wash any affected areas with large quantities of water. Notify your instructor and seek a medical evaluation of the chemical burn.

To test a detergent for phosphate, add about 60 mg of the detergent (a quantity about the size of a grain of rice) in a micro test tube and add 8 drops of 6 M nitric acid (HNO_3). Some detergents contain carbonates that cause foaming when HNO_3 is added. If this occurs, add HNO_3 dropwise until foaming ceases; then add 8 drops more.) If the detergent is a liquid, then use about 15-20 drops of the detergent for this test. Using a glass stirring rod, mix the solution thoroughly. Transfer 5 or 6 drops of the solution to another micro test tube. Add 2 drops of 0.2 M ammonium molybdate solution, and warm gently in boiling water for about 5 minutes. A yellow precipitate, which may form slowly, confirms the presence of phosphate in the detergent. Perform this test for each detergent provided.

Report Sheet

_____   _____   _____
Name   Instructor/Section   Date

Part A—Solubility of Cottonseed Oil

Solvent	Observations
Water	_____
Ethanol	_____
Hexane	_____
1,4-Dioxane	_____

Part B—Unsaturation of Triacylglycerols

Substance	Number of drops of Br_2 solution	Conclusions
Cottonseed oil	_____	_____
Fat	_____	_____

Part C—Properties of Soaps and Detergents

Test	Your soap	Commercial soap	Detergent
Litmus	_____	_____	_____
pH paper	_____	_____	_____
Emulsifying properties	_____	_____	_____
Calcium ion	_____	_____	_____
Magnesium ion	_____	_____	_____
Iron(III) ion	_____	_____	_____

Part D—Test for Phosphate Ion

Detergent	Results of phosphate test	Phosphate content from label
Fab	_____	_____
_____	_____	_____
_____	_____	_____
_____	_____	_____

Answer the questions on the following page.

QUESTIONS

1. Which of the solvents tested would be best to remove an oil stain from clothing?

2. How many grams of bromine are required to react completely with 1 mole of a triacylglycerol containing only linolenic acid?

3. Why was a concentrated solution of sodium chloride used to precipitate the soap?

4. What reaction occurs when an aqueous solution of calcium ion is added to a soap solution?

5. Does the solubility of the metal salts of carboxylic acids and alkyl sulfates parallel the solubility of the metal salts of CO_3^{2-} and SO_4^{2-}?

6. Arsenate ion, AsO_4^{3-}, will also react with ammonium molybdate in the same way as phosphate. Explain why. What ion is produced?

34 Analysis for Vitamin C

APPARATUS

Burner, rubber tubing, matches, beakers, Erlenmeyer flasks, clamp, iron ring, wire gauze, 25 mL buret, 1.00 mL volumetric pipet, spatula, buret clamps.

REAGENTS

Acetic acid-metaphosphoric acid solution, ascorbic acid standard solution (approximately 0.0140 M), 2,6-dichlorophenolindophenol solution (1.0×10^{-3} M), powdered drink mixes ("Tang" and "Country Time"), solid unknown samples of ascorbic acid.

INTRODUCTION

Vitamin C is a component of fresh fruits and vegetable. Vitamin C is decomposed readily. It is oxidized by the oxygen in the air and by other oxidizing agents. Prolonged heating also destroys this compound. For these reasons, foods containing vitamin C must be stored and prepared carefully if the vitamin C content is to be maintained. In 1974, the Food and Nutrition Board of the National Academy of Sciences set the recommended daily allowances for vitamin C (ascorbic acid) as follows.

Infants	35 mg
Adolescents	40 mg
Adults	45 mg
Pregnant women	60 mg
Lactating women	80 mg

One method for the determination of ascorbic acid is an oxidation-reduction titration. Ascorbic acid reacts quantitatively and rapidly with 2,6-dichlorophenolindophenol (DCPIP). DCPIP is blue in basic solution and pink in acid. The deep blue solution of DCPIP used for titration becomes colorless due to reduction by ascorbic acid. After all the ascorbic acid in the titration flask has reacted, the subsequent addition of the blue basic DCPIP will turn pink in the titration flask since it contains acid. A mixture containing acetic acid and metaphosphoric acid is added to inhibit the oxidation of the ascorbic acid by the oxygen in the air.

ascorbic acid	DCPIP	dehydroascorbic acid	reduced DCPIP
colorless	blue	colorless	colorless

In this experiment the amount of ascorbic acid will be determined in a powdered drink mix and in an unknown assigned by the instructor.

PROCEDURE

Study Section J, "Volumetric Analysis," on pages 18-22 before starting this experiment.

Aqueous solutions of 2,6-dichlorophenolindophenol (DCPIP) decompose slowly upon standing. Thus, the solution is standardized against a known quantity of ascorbic acid just prior to determining the ascorbic acid content in the unknown. The standardization procedure is the same as that for the titration of the unknown. The ascorbic acid is titrated with the blue solution of DCPIP until the last drop of DCPIP changes the colorless ascorbic acid solution reddish pink.

Part A—Standardization of DCPIP Solution

Place about 20 mL of the standard solution of ascorbic acid in a clean, dry labeled 125 mL Erlenmeyer flask. Record the exact concentration from the label of the original bottle. Keep the flask that contains the standard solution stoppered between uses. Obtain about 250 mL of the DCPIP solution in another labeled stoppered flask.
Carefully clean the buret with detergent. Rinse it thoroughly with tap water and distilled water. Finally, rinse it twice with 3 mL portions of the 2,6-dichlorophenolindophenol (DCPIP) solution, and fill it with the same solution.
Obtain a 1 mL pipet from the instructor. Clean it and rinse it with distilled water and with the ascorbic acid solution. Carefully pipet exactly 1 mL of the standard ascorbic acid solution into each of two clean but not necessarily dry 250 mL Erlenmeyer flasks. Add to each flask about 100 mL of distilled water and about 1 mL of the acetic acid-metaphosphoric acid solution. Titrate the contents of each flask with the DCPIP solution until the light pink color persists permanently. Record your data on the Report Sheet. If the difference between the two titration volumes is more than 0.20 mL then titrate a third sample.

Part B—Determination of Ascorbic Acid in a Powdered Drink Mix

Determine the ascorbic acid content of one lemonade powder or orange juice powder (containing Nutra Sweet) by titrating two precisely weighed samples. For each sample use the following approximate size.

Powdered mix	Sample size
"Country Time"	0.7 g
"Tang"	0.1 g

Add 75 mL of distilled water and 1 mL of acetic acid-metaphosphoric acid solution. Titrate the contents of each flask with your DCPIP solution. Record your data on the Report Sheet. Perform a third titration if the results do not agree within 0.04 mg of ascorbic acid.

Part C—Determination of Ascorbic Acid in Solid Unknown

Determine the vitamin C content of the unknown obtained from the instructor. Titrate at least two precisely weighed 1 g samples. The samples contain sucrose and dissolve slowly. Do not heat the sample to accelerate its dissolving. Perform a third titration if the results do not agree within 0.04 mg of ascorbic acid.

Part D—Titration of Heated Ascorbic Acid Solution

Finally test the sensitivity of ascorbic acid to heat. Pipet 1 mL of the standard solution of ascorbic acid into a 250 mL flask containing 100 mL of distilled water. Heat for 10 minutes. Cool the flask, add 1 mL of acetic acid-metaphosphoric acid solution, and titrate with your DCPIP solution. Compare the result with that obtained with 1 mL of the unboiled solution.

CALCULATIONS

The molarity of the standardized DCPIP solution M_D, is calculated from the equation

$$M_d = \frac{1.00 \text{ mL} \times M_A}{V_D} \tag{1}$$

where M_A is the molarity of the ascorbic acid solution and V_D is the volume in mL of the DCPIP solution used for titrating 1.00 mL of the ascorbic acid solution.

The number of moles of DCPIP is the product of the volume in mL of DCPIP times the molarity of the DCPIP, M_D.

$$\text{Moles DCPIP} = V_D \times M_D \tag{2}$$

The number of moles of ascorbic acid in a sample is equal to the number of moles of DCPIP used to titrate the sample. The weight of ascorbic acid is the number of moles multiplied by its molecular weight.

Weight ascorbic acid = moles ascorbic acid × molecular weight ascorbic acid   (3)

The weight of ascorbic acid per gram of sample is obtained by dividing the weight of ascorbic acid by the total weight of the sample in grams.

Report Sheet

_____ _____ _____
Name   Instructor/Section   Date

Part A—Standardization of DCPIP Solution

	Trial 1	Trial 2	Trial 3
Volume of ascorbic acid solution	_____	_____	_____
Molarity of ascorbic acid solution	_____	_____	_____
Initial buret reading (mL)	_____	_____	_____
Final buret reading (mL)	_____	_____	_____
Volume of DCPIP solution used (mL)	_____	_____	_____
Molarity of DCPIP	_____	_____	_____
Average molarity of DCPIP	_____		

Part B—Determination of Ascorbic Acid in Powdered Drink Mix

Sample _____

	Trial 1	Trial 2	Trial 3
Weight of container + sample	_____	_____	_____
Weight of container	_____	_____	_____
Weight of sample	_____	_____	_____
Molarity of DCPIP	_____	_____	_____
Initial buret reading (mL)	_____	_____	_____
Final buret reading (mL)	_____	_____	_____
Volume of DCPIP	_____	_____	_____
Milligrams of ascorbic acid per gram of sample	_____	_____	_____
Average mg ascorbic acid/g of sample	_____		

Analysis for Vitamin C

Part C—Determination of Ascorbic Acid in the Solid Unknown

Unknown number _____

	Trial 1	Trial 2	Trial 3
Weight of container + sample	_____	_____	_____
Weight of container	_____	_____	_____
Weight of sample	_____	_____	_____
Molarity of DCPIP	_____	_____	_____
Initial buret reading (mL)	_____	_____	_____
Final buret reading (mL)	_____	_____	_____
Volume of DCPIP	_____	_____	_____
Milligrams of ascorbic acid per gram of sample	_____	_____	_____
Average mg ascorbic acid/g of sample	_____		

Part D—Titration of Heated Ascorbic Acid Solution

	Trial 1	Trial 2	Trial 3
Volume of ascorbic acid solution	_____	_____	_____
Initial buret reading (mL)	_____	_____	_____
Final buret reading (mL)	_____	_____	_____
Volume of DCPIP	_____	_____	_____
Molarity of heated ascorbic acid solution	_____	_____	_____
Average molarity of heated ascorbic acid	_____		

Answer the questions on the following page.

262

QUESTIONS

1. Several tomatoes are bisected. One half of each tomato is made into stewed tomatoes by heating. The other half of each tomato is used to prepare tomato juice without heating. Which preparation would contain the most vitamin C? Explain your answer.

2. What difficulties would you expect to encounter if the method described in this experiment were used to titrate the vitamin C in tomato juice?

3. Why might canned orange juice contain more vitamin C per liter than freshly squeezed orange juice?

4. Assume the solution of ascorbic acid used for standardizing the DCPIP solution was very old. Would the use of old standard ascorbic acid solution cause your result for your unknown to be artificially high or low? Explain your answer.

5. Why should you not heat the solution containing the solid unknown in order to make it dissolve faster?

EXPERIMENT
35

Analysis of Blood Glucose

APPARATUS

Six 16 × 150 mm test tubes, six 10 × 75 mm test tubes, two 5.00 mL volumetric pipets, test tube rack, spectrophotometer, and cuvettes.

REAGENTS

Solutions of 25, 50, 75, and 100 mg glucose per 100 mL, Benedict's solution diluted 1:8, tissue paper for cleaning cuvettes

INTRODUCTION

The normal concentration of glucose in the blood should be in the 65–95 mg/100mL range. The units used in clinical analysis are mg/dL. The blood glucose level may decrease temporarily during strenuous exercise, because the glucose may not be replenished rapidly enough from liver glycogen or by gluconeogenesis. A condition of low blood glucose is called hypoglycemia. It is characterized by a rapid heartbeat, general weakness, trembling, perspiration, whitening of the skin, and loss of consciousness. The loss of consciousness is due to the deprivation of brain cells of the necessary glucose.

A condition of high blood glucose is called hyperglycemia. The digestion of carbohydrates may result in absorption of glucose into the blood faster than glycogen can be formed by the process of glycogenesis. As the blood glucose level increases, the body also transforms glucose into fat and stores the fat as adipose tissue. When blood glucose reaches the level of 140–160 mg per 100 mL of blood, neither glycogen nor fat can be formed rapidly enough to decrease the glucose level. The condition in which glucose is then excreted by the kidneys and is eliminated in the urine is called glucosuria.

Tests for the concentration of glucose in blood and urine are done in clinical laboratories. Modern methods use automated analytical procedures that rapidly produce colored products that can be analyzed by spectrophotometers. The concentration of the colored species is determined by measuring its absorbance at a specific wavelength. The absorbance, A, is directly proportional to its concentration.

$$A = \varepsilon bc \qquad (1)$$

where ε is the molar absorptivity (a constant for the species at that wavelength), b is the length of the cell, and c is the molarity of the species. Thus, for a specific cell (usually $b = 1$ cm) and for a selected wavelength, the larger the absorbance, the higher is the concentration of the absorbing species. As a consequence, one can determine the absorbance for known concentrations of a species and use these values to determine the proportionality constant. Using that constant, the concentration of an unknown solution can be determined after measuring the absorbance of the solution.

In this experiment, a slightly different approach is used that employs a test reagent with which you are familiar—Benedict's solution. Furthermore, this experiment uses specifically prepared aqueous glucose solutions rather than body fluids. Glucose reacts with the Cu^{2+} complex ion of Benedict's solution to give solid Cu_2O. As a result of this reaction the concentration of the Cu^{2+} ion decreases. The resulting decrease in the absorbance of the solution at the wavelength of maximum intensity for the Cu^{2+} is directly proportional to the glucose concentration.

In this experiment you will determine the linear relationship between absorbance and concentration using solutions generated by the reaction of various known concentrations of glucose with Benedict's solution. Using this calibration line, you can then determine the concentration of a solution of glucose of unknown concentration. Because all known solutions and the unknown solution are treated identically, it is not necessary to determine the molar absorptivity of the Cu^{2+} ion. The Benedict's solution used is of lower concentration than the standard solution in order to provide solutions that do not exceed the range of the spectrophotometer used in your laboratory.

PROCEDURE

Study Section J, "Volumetric Analysis," on page 19 showing the proper use of the pipet. Study the instructions given in Section L, "Spectrophotometry," on pages 24-25 of this manual for the proper use of the spectrophotometer. The instrument contains expensive electronic and optical components that must be handled carefully to prevent damage.

Do not make any adjustments that you do not understand.

In this experiment, the spectrophotometer should be set at 730 nanometers and allowed to warm up for 20 minutes. At this wavelength the absorbance due to Cu^{2+} is the strongest and thus the results will be most accurate.

Obtain approximately 10 mL of each of the four standard glucose solutions in labeled clean dry vessels such as 16 × 150 mm test tubes. Obtain approximately 40 mL of dilute Benedict's solution. Obtain an unknown solution from your instructor.

Prepare six 16 × 150 mm test tubes for this experiment by washing with distilled water and drying. Number the tubes 1 through 6 and pipet 5.00 mL of the dilute Benedict's solution into each tube. Use a different pipet for the addition of the second component to the test tube. Into tube 1 pipet 5.00 mL of distilled water. Into tubes 2 through 5, pipet 5.00 mL of the different standard glucose solutions. Into tube 6, pipet 5.00 mL of the unknown glucose solution. The pipet should be rinsed with distilled water before using it with a solution of a different concentration. Then rinse it with a small amount of the solution to be added.

Gently agitate each test tube to ensure complete mixing. Place all tubes in a 400 mL beaker containing 150 mL of hot water maintained at a gentle boil by a Bunsen burner. After 30 minutes, remove the test tubes and place them in a test tube rack to cool.

If the Cu_2O formed in the reaction settles to the bottom of the test tubes it will be possible to decant the solution. However, it may be necessary to place a portion of the contents of the test tubes into 10 × 75 mm test tubes and centrifuge them to pack the Cu_2O at the bottom.

Obtain two cuvettes from your instructor. Wash and rinse both of them. Add distilled water to one cuvette and set it aside as your reference cell. The other cuvette is your sample cell. Rinse it two or three times with small amounts of the solution from tube 1. Then fill the cuvette two-thirds full with the solution from tube 1.

Insert the cuvette with distilled water in the spectrophotometer and set the %T to 100%. Remove the cuvette and insert the sample cuvette containing the contents of tube 1. Record the absorbance on the Report Sheet.

Determine the absorbance of each cool solution. Rinse the sample cell two or three times with small amounts of each new solution before filling the tube for the absorbance determination. This procedure ensures that the concentration of the solution in the cuvette is the same as in the sample solution and is not contaminated by solutions previously placed in the cuvette.

Record the absorbance of each solution on your Report Sheet. Prepare a calibration curve for the determination of glucose. Plot the absorbance on the vertical axis and the original concentration of glucose on the horizontal axis. Because the concentration of the Cu^{2+} decreases with increasing glucose concentration the line has a negative slope. The intercept on the vertical axis corresponds to 0 mg glucose per 100 mL for tube 1. Draw the best straight line through this intercept and the four other points for specified concentrations of glucose. A sample calibration line is shown in Figure 35-1. Use the line to determine the concentration of the glucose solution of unknown concentration.

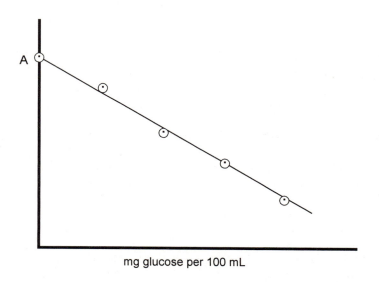

mg glucose per 100 mL

Figure 35-1. Sample calibration line for determination of glucose concentration.

Report Sheet _____   _____   _____

                Name   **Instructor/Section**   **Date**

Tube	Initial concentration of glucose	Absorbance at 730 nm
1	_____	_____
2	_____	_____
3	_____	_____
4	_____	_____
5	_____	_____
6	Unknown	_____

Include the graph showing the determination of the concentration of glucose of the unknown. Answer the questions on the next page.

QUESTIONS

1. What would be the effect of rinsing the cuvette with distilled water prior to introducing a new solution? Assume a few drops of distilled water remained in the cuvette.

2. Based on your calibration curve, what concentration of glucose would give zero absorbance?

3. How could the degree of the fit of the calibration curve be improved by a change in the experimental conditions?

4. Would the determination of the concentration of the unknown solution be affected if the Benedict's solution were inadvertently prepared to give a solution of 5% greater concentration of Cu^{2+}?

5. Assume that the solution with 25 mg of glucose per 100 mL was incorrectly prepared and the concentration is actually 35 mg of glucose per 100 mL. How would this error be detected?

EXPERIMENT

36   Bioinorganic Chemistry

APPARATUS

10 × 75 mm test tubes, gummed labels, test tube rack, medicine droppers, spectrophotometer, and cuvettes.

REAGENTS

0.025 M solutions of $Ca(NO_3)_2$, $Co(NO_3)_2$, $Cr(NO_3)_3$, $Cu(NO_3)_2$, $FeSO_4$, $Mn(NO_3)_2$, and $Ni(NO_3)_2$; 1 M $K_2C_2O_4$, 1 M NH_3, 1 M ethylenediamine, 0.5 M 1,10-phenanthroline, 0.25 M $Ni(NO_3)_2$, and 0.20 M $Co(NO_3)_2$.

INTRODUCTION

Trace quantities of metal ions play important roles in biological systems. However, the ions are not free, but are surrounded by complex organic molecules containing several functional groups that are coordinated to the metal ion. The groups coordinated to the metal ion, called ligands, supply one or more electron pairs to the metal ion. Simple ligands such as water or ammonia that supply one electron pair as well as ions such as chloride ion or cyanide ion are called monodentate ligands. More complex ligands may contain two or more electron pairs. A bidentate ligand supplies two electron pairs and forms two bonds to a single metal ion.

In biological systems a single organic molecule may supply from two to four or more electron pairs. In hemoglobin, the heme subunit provides four electron pairs from four nitrogen atoms. A fifth pair of electrons is provided by an amino acid in the globin portion of hemoglobin. The sixth coordinate position of the central iron atom is occupied by oxygen, which is transported for use by the organism. Other important bioinorganic ions include the complex of magnesium in chlorophyll, cobalt in vitamin B_{12}, molybdenum in bacterial enzymes that fix nitrogen, and iron in cytochromes in the electron transport chain.

The activity of many enzymes depends on specific metal ions. For example, carboxypeptidase contains zinc(II) bonded to two nitrogen atoms and one oxygen atom of three amino acids of the protein. The fourth coordinate position is free to catalyze chemical reactions by coordination of the substrate. Substitution of zinc by cobalt(II) does not affect the reactivity of the enzyme. However, mercury(II) completely eliminates the activity of the enzyme.

In this experiment we will examine the properties of complex ions formed from metal ions coordinated to simple ligands. The structure of the complex ion depends on the coordination properties of the metal ion and the number of sites of the ligands that can supply electron pairs. For a monodentate ligand such as cyanide ion (CN^-) the number of ligands depends on the metal ion, as illustrated by $Fe(CN)_6^{3-}$, $Cu(CN)_4^{2-}$, and $Ag(CN)_2^-$. The number of monodentate ligands that coordinate to a metal ion is its coordination number. Coordination numbers from 2 to 8 are known, but 4 and 6 are the most common. With the exception of copper, whose coordination number is 4, the remaining transition metal ions used in this experiment have a coordination number of 6.

The properties of the complex ion depend on the ligand. For example, the aquo complex ion of copper(II), which is represented by $Cu(H_2O)_4^{2+}$, is light blue but the ammonia complex, represented by $Cu(NH_3)_4^{2+}$, is a substantially deeper blue-violet. In this experiment we will determine the color of several metal ions with three ligands—water, ammonia, and the bidentate ligand oxalate.

oxalate ion

The strength of the binding of ligands reflects the electron donor properties of the ligand. In the case of Cu^{2+}, ammonia complexes more strongly than water does. Thus, the position of the following equilibrium is strongly to the right.

$$Cu(H_2O)_4^{2+} + 4\ NH_3 \rightleftharpoons Cu(NH_3)_4^{2+} + 4\ H_2O$$

In this experiment we will examine the position of similar equilibria by observing the color that predominates in a solution that contains two competing ligands.

Bidentate ligands such as ethylenediamine, abbreviated "en", and 1,10-phenanthroline, abbreviated "phen", each can supply two electron pairs.

$NH_2\!-\!CH_2CH_2\!-\!NH_2$

ethylenediamine   1,10-phenanthroline

Thus, ethylenediamine forms a 2:1 complex with Cu^{2+} (coordination number 4) because each molecule supplies two electron pairs to the metal ion. The abbreviation for the complex ion is $Cu(en)_2^{2+}$. 1,10-Phenanthroline forms a 3:1 complex with Fe^{2+} (coordination number 6) because each molecule supplies two electron pairs to the metal ion. The abbreviation for the complex ion is $Fe(phen)_3^{2+}$.

The concentration of a complex ion can be determined spectrophotometrically. The absorbance, A, of a single species at a specific wavelength is directly proportional to its concentration.

$$A = \varepsilon bc \qquad\qquad (1)$$

where ε is the molar absorptivity (a constant for the species at that wavelength), b is the length of the cell, and c is the molarity of the species. Thus, for a specific cell (usually $b = 1$ cm) and for a selected wavelength, the larger the absorbance, the higher is the concentration of the absorbing species. As a consequence, one can determine the absorbance for a known concentration of a species and determine the proportionality constant. Using that constant, the concentration of an unknown solution can be determined by measuring the absorbance of the solution.

The absorbance of a mixture of components X and Y, at a particular wavelength is the sum of the absorbances of each species in the solution, as given by equation (2).

$$A' = \varepsilon_x' \, b \, [X] + \varepsilon_y' \, b \, [Y] \qquad\qquad (2)$$

where the prime refers to measurements made at a wavelength λ'. The proportionality constants ε_x' and ε_y' are the molar absorptivities of the species X and Y at that wavelength.

To determine the individual concentrations of X and Y, it is necessary to obtain spectrophotometric information at a second wavelength λ'' and know the molar absorptivity of both individual components at that wavelength. The absorbance of a mixture of components X and Y at this second wavelength is the sum of the absorbances of each species in the solution as given by equation (3).

$$A'' = \varepsilon_x'' \, b \, [X] + \varepsilon_y'' \, b \, [Y] \qquad\qquad (3)$$

where the double prime refers to measurements made at a wavelength λ''. The proportionality constants ε_x'' and ε_y'' are for the molar absorptivities for the species X and Y at that wavelength.

To analyze a mixture of two components, it is necessary to measure the absorbance at two wavelengths and to know ε for each component at each wavelength. The quantities [X] and [Y] may then be calculated by simultaneous solution of the equations (2) and (3). It is customary to select one wavelength where the absorbance of species X is small compared to that of species Y and a second wavelength where the absorbance of species Y is small compared to that of species X. Such experimental circumstances are illustrated in Figure 41-1. The X species has a maximum absorption at λ', and the Y species has a maximum absorption at λ''. A mixture of X and Y is shown on the composite curve.

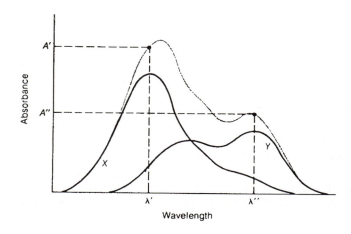

Figure 36-1. Absorbance of pure substances and a composite mixture.

PROCEDURE

For Part E, study the instructions given in Section L, "Spectrophotometry," on pages 24-25 of this manual for the proper use of the spectrophotometer. The instrument contains expensive electronic and optical components that must be carefully handled to prevent damage.

Part A—Color of Aquo Ions

Place 20 drops (1 mL) of 0.025 M aqueous solutions of Ca^{2+}, Co^{2+}, Cr^{3+}, Cu^{2+}, Fe^{2+}, Mn^{2+}, and Ni^{2+} in separate labeled 10 × 75 mm test tubes. Record the color of the aquo ion on the Report Sheet. Note that the anions of the salts used are colorless and thus the color of each solution corresponds to that of the cation.

Part B—Coordination of Metal Ions by Oxalate Ion

To the tubes containing Ca^{2+}, Co^{2+}, Cr^{3+}, and Cu^{2+}, add 5 drops of 1 M $K_2C_2O_4$. Describe the color of the resulting solution in each test tube. Record the color on the Report Sheet and write the formula of the complex ion product.

Part C—Coordination of Metal Ions by 1,10-Phenanthroline

Place 20 drops (1 mL) of 0.025 M aqueous solutions of Co^{2+} and Cu^{2+} in separate labeled 10 × 75 mm test tubes. To these tubes and the two tubes containing Fe^{2+} and Mn^{2+} from Part A, add 5 drops of 0.5 M 1,10-phenanthroline. Record the color on the Report Sheet and write the formula of the complex ion based on the coordination number of metal ion.

Part D—Binding Strengths of Ligands toward Ni^{2+}

Place 20 drops of 0.25 M Ni^{2+} in each of four separate 10 × 75 mm test tubes labeled 1 through 4. Then add the following reagents to the test tubes as indicated.

Tube 1 Add 5 drops of 1 M NH_3. Record the color of the resulting solution.

Tube 2 Add 5 drops of 1 M ethylenediamine. Record the color of the resulting solution and compare it to both the aquo complex ion and the color of the ammonia complex ion of Tube 1.

Tube 3 Add 5 drops of 0.5 M 1,10-phenanthroline. Record the color of the resulting solution.

Tube 4 Add 5 drops of 1 M $K_2C_2O_4$. Record the color of the resulting solution.

Now add a second reagent to the indicated test tubes as follows.

Tube 1 Add 5 drops of 1 M ethylenediamine. Record the color of the resulting solution. Based on this information, determine the ligand strength of ammonia relative to ethylenediamine.

Tube 2 Add 5 drops of 0.5 M 1,10-phenanthroline. Record the color of the resulting solution. Based on this information, determine the ligand strength of ethylenediamine relative to 1,10-phenanthroline.

Tube 3 Add 5 drops of 1 M NH_3. Record the color of the resulting solution. Based on this information, determine the ligand strength of ammonia relative to 1,10-phenanthroline.

Tube 4 Add 5 drops of 1 M NH_3. Record the color of the resulting solution. Based on this information, determine the ligand strength of ammonia relative to $C_2O_4^{2-}$.

At this point, you should be able to rank the binding strengths of the ligands H_2O, NH_3, ethylenediamine, 1,10-phenanthroline, and oxalate toward Ni^{2+}.

Part E—Analysis of a Mixture of Metal Ions

Obtain and clean two cuvettes. Rinse them with distilled water. Obtain approximately 10 mL samples of 0.20 M $Co(NO_3)_2$ and 0.25 M $Ni(NO_3)_2$ and an unknown mixture of these ions from your instructor. Fill one cuvette with distilled water to within 2 cm of the top. Use the second cuvette for the solutions of the metal ions. For each ion, wash out the cuvette with distilled water followed by three successive washings with approximately 1 mL of the metal ion to be used. This washing process must be repeated each time a different metal ion solution is to be studied. Discard the wash solutions in a beaker.

The spectrophotometer, which should be warmed up for 20 minutes, is used to determine the absorbance of the solutions at each of two wavelengths. Set the wavelength control at 520 nm. Insert the cuvette with distilled water and set the %T to 100%. Remove the cuvette and insert the cuvette containing $Co(NO_3)_2$. Record the absorbance on the Report Sheet. Then replace the sample with $Ni(NO_3)_2$. Finally replace the sample by the unknown mixture. Record the absorbance for each sample on the Report Sheet.

Set the wavelength control at 720 nm. Insert the cuvette with distilled water and set the %T to 100%. Remove the cuvette and insert the cuvette containing $Co(NO_3)_2$. Record the absorbance on the Report Sheet. Then successively replace the sample with $Ni(NO_3)_2$, followed by the unknown mixture, recording each absorbance on the Report Sheet.

Report Sheet _____   _____   _____
Name   Instructor/Section   Date

Part A—Color of Aquo Ions

Metal Ion	Color	Metal Ion	Color
Ca^{2+}	_____	Co^{2+}	_____
Cr^{3+}	_____	Cu^{2+}	_____
Fe^{2+}	_____	Mn^{2+}	_____
Ni^{2+}	_____		

Part B—Complex Product with Oxalate Ion

Metal Ion	Color	Formula of Product
Ca^{2+}	_____	_____
Co^{2+}	_____	_____
Cr^{3+}	_____	_____
Cu^{2+}	_____	_____

Part C—1,10-Phenanthroline Complex Ions

Metal Ion	Color	Formula of Complex Ion
Co^{2+}	_____	_____
Cu^{2+}	_____	_____
Fe^{2+}	_____	_____
Mn^{2+}	_____	_____

Part D—Ligand Strength Toward Ni^{2+}

Tube	First Reagent	Color	Second Reagent	Color
1.	NH_3	_____	ethylenediamine	_____
2.	ethylenediamine	_____	1,10-phenanthroline	_____
3.	1,10-phenanthroline	_____	NH_3	_____
4.	oxalate	_____	NH_3	_____

Bioinorganic Chemistry

Part E—Composition of a Mixture

Sample	Absorbance at 520 nm	Absorbance at 720 nm
Ni^{2+}	_____	_____
Co^{2+}	_____	_____
Unknown	_____	_____

Metal Ion	Molar Absorptivity at 520 nm	Molar Absorptivity at 720 nm
Ni^{2+}	_____	_____
Co^{2+}	_____	_____

Molarity of Ni^{2+} in the unknown _____

Molarity of Co^{2+} in the unknown _____

Attach a sheet showing your calculations and the answers to the questions.

QUESTIONS

1. Compare the ligand strength of water versus ammonia. How is it related to the basicity of the two ligands?

2. The ligand strength of each nitrogen atom of ethylenediamine is comparable to that of ammonia. Explain why ethylenediamine is a stronger ligand than ammonia. (Hint, consider the balanced equation and recall the effect of entropy on chemical reactions.)

3. Suggest a reason why the dianion of adipic acid does not form bidentate complex ions.

4. How could a mixture of three metal ions be analyzed? What additional experiments would have to be done?

5. The absorbances of 0.08 Cu^{2+} at 520 and 720 nm are 0.012 and 0.64, respectively. Can a mixture of Cu^{2+} and Ni^{2+} be conveniently analyzed? Can a mixture of Cu^{2+} and Co^{2+} be conveniently analyzed? Why or why not?

EXPERIMENT
37   Chromatography

APPARATUS

Capillary tubes, Whatman No. 1 filter paper(12 × 25 cm), 1000 mL beaker, aluminum foil, ninhydrin sprayer, oven, stapler.

REAGENTS

2% aqueous ammonia, 2-propanol, 1.5% hydrochloric acid, 0.05 M solutions of aspartic acid, glycine, leucine, tyrosine in 1.5% HCl, unknown mixtures of amino acids, 2% ninhydrin in ethanol.

INTRODUCTION

In addition to recrystallization and distillation, a third technique, called chromatography, can be used to separate the components of a homogeneous mixture. Initially chromatography was used to separate colored substances; hence, the name "chromatography" (Greek *chroma,* color). The colors of the components of the mixture are observed as they separate. Chromatography can be also used with colorless substances if they fluoresce when exposed to ultraviolet light or if they react with a second reagent to produce colored products.

Separation of substances by chromatography depends on the differences between the adsorptive characteristics of the substances with respect to a stationary reference material such as paper. Usually, a liquid solution containing a mixture of substances is passed through a system containing such a reference material called the stationary phase. The components of the mixture are adsorbed on the stationary phase. Continued passage of a solvent in the stationary phase dissolves the adsorbed compounds and moves them to a new adsorption site. Thus, by a process of reversible adsorption and dissolution the components of a mixture will be moved through the stationary phase. Each component of the mixture moves at its own rate. Thus, after a given time interval each component has moved a different distance across the stationary phase.

In this experiment, amino acids will be placed on a sheet of paper and the solvent allowed to move along the paper by capillary action for a given period of time. The paper is the stationary phase, and the amino acids will move along the paper at rates that depend on their structures. For a given amino acid, the ratio of the distance traveled relative to that traveled by the solvent is the R_f value for the amino acid.

$$R_f = \frac{\text{distance traveled by compound}}{\text{distance traveled by solvent}}$$

Since amino acids are colorless, identification of their positions at the end of the experiment is necessary. Ninhydrin will be used to develop a spot of color at the point to which each amino acid has moved.

PROCEDURE

CAUTION: Ninhydrin will strongly color the amino acids in your skin. Avoid spilling ninhydrin on yourself.

Prepare a solution of 10 mL of 2% aqueous ammonia in 20 mL of 2-propanol. The solution should be placed in a 1000 mL beaker and covered with aluminum foil.

Obtain a few drops of 0.05 M solutions of each of the amino acids (aspartic acid, glycine, leucine, and tyrosine) in 1.5% hydrochloric acid. Place the amino acids in labeled test tubes, and place a capillary tube in each tube. Obtain a few drops of an unknown mixture from your instructor.

Make a light pencil mark along the long axis of a sheet of Whatman No. 1 filter paper (12 × 25 cm) at a point 2 cm from the edge. Along this line place ten x marks at equal intervals. Place the first mark 2 cm from the short edge of the paper (Figure 46-1). A distance of 2 cm between marks will allow proper placement of all the x marks. Mark the edge of the paper below each x with an identifying mark. The unknown mixture mark should be placed between groups of two amino acids. Repeat your order of amino acids a second time to obtain two spots for each amino acid and the unknown.

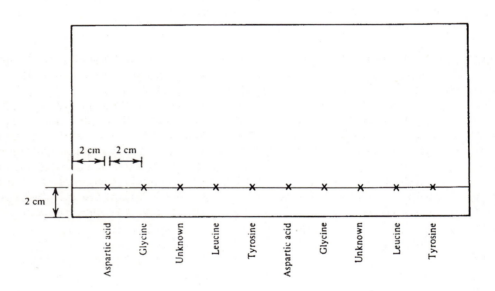

Figure 37-1. Preparation of Chromatographic Sheet.

Place a small drop of each solution at the two positions on the line on the filter paper. It is advisable to practice with ordinary filter paper prior to preparing your sheet for chromatography. The spot of solution on the paper should be no larger than 2 or 3 mm. After placing all samples on the paper, allow the paper to dry in the air for 10 minutes.

Roll the paper into a cylinder such that the line is on the bottom outside the cylinder. Staple the ends together to hold the cylinder in shape. Place the staples about 4 cm from each edge of the cylinder. Do not allow the edges of the paper to touch when the staples are put in. A small gap in the cylinder is necessary (Figure 46-2).

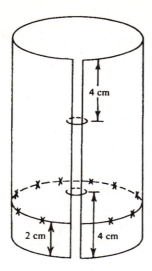

Figure 46-2. Construction of Chromatographic Cylinder.

Place the cylinder, base line down, in the beaker of solvent and cover the beaker tightly with aluminum foil. Avoid splashing the solvent on the paper. Make sure that the paper does not touch the sides of the beaker.

After at least 90 minutes, remove the paper cylinder and mark the edge of the wet part of the paper and the depth to which the paper was submerged in the solvent. Place the rolled chromatogram upside down on a piece of paper to dry. After the cylinder is essentially dry, remove the staples and hang it in a hood. Have your instructor spray the paper with a solution of ninhydrin. After the spray solution dries, put the paper in an oven at 100°C for 10-15 minutes. Note the spots, and circle each one. Measure the distances from the x to each spot and the distance traveled by the solvent. The distance traveled by the solvent is the distance from the liquid level in the beaker to the edge of the wet portion of the paper. Calculate the R_f values and determine the identity of your unknown.

Report Sheet

Name _____ **Instructor/Section** _____ **Date** _____

Summary of results

Amino acid	Distance traveled by an Amino acid	Distance traveled by solvent	R_f value
Aspartic acid	_____	_____	_____
Glycine	_____	_____	_____
Leucine	_____	_____	_____
Tyrosine	_____	_____	_____
Unknown	_____	_____	_____

Components of unknown mixture: _____

Answer the questions on the next page.

QUESTIONS

1. How might it be possible to determine quantitatively the composition of an amino acid mixture?

2. If two amino acids have the same R_f values in 2-propanol, how might they be separated?

3. Draw the structure of the amino acid in your unknown.

EXPERIMENT

38   Catalysis by an Enzyme

APPARATUS

Two 250 mL beakers, 10 mL graduated cylinder, two 2 mL volumetric pipets, 25 × 200 mm test tubes, rubber tubing, glass tubing, clock with second hand, thermometer, mortar and pestle.

REAGENTS

3% hydrogen peroxide, 0.1 M copper(II) sulfate, 0.1 M hydrochloric acid, 0.1 M sodium hydroxide, catalase solution, ice.

INTRODUCTION

Reactions that occur in living systems involve common organic functional groups such as hydroxyl, carbonyl, and amino. However, biological molecules are very large and complex. Many reactions of living systems (*in vivo*) can also be run in laboratory glassware (*in vitro*) to yield the same products. For the reaction in vitro to occur at the same rate as in vivo, the conditions of temperature, concentration, pH, and catalysis must be identical.

Biological catalysts are protein molecules called enzymes. Without enzymes, biological reactions would require either high temperatures or long reaction times to obtain reasonable yields. Enzymes isolated from living systems can be used *in vitro* to catalyze a reaction that otherwise occurs only *in vivo*.

The reactant in a biological reaction is the substrate. Enzymes catalyze specific reactions of certain substrates. Some enzymes catalyze one type of reaction with a variety of substrates such as hydrolysis of esters. Other enzymes are so specific that only a single reaction of a single substrate is catalyzed.

The three-dimensional structures of enzymes feature clefts or crevices that allow the substrate and the enzyme to fit together. The resultant complex has the reacting site of the substrate in close proximity to the site of catalytic activity on the enzyme.

The forces that maintain the shape of the enzyme include salt bridges, hydrogen bonds, disulfide bonds, and hydrophobic interactions. Any chemicals that can change these forces within the enzyme will eliminate the ability of that enzyme to catalyze a reaction. An altered and inactive enzyme is denatured. Heat will denature all enzymes. The effects of metal ion concentration and pH on the activity of an enzyme depend on the particular enzyme.

Many enzymes are affected by chemicals known as inhibitors. Competitive inhibitors occupy the active site on the enzyme, thereby blocking the substrate. Noncompetitive inhibitors bind at some other site in the enzyme, causing a conformational change near the active site. Both types of inhibitors eliminate the catalytic power of the enzyme.

Catalysis by an Enzyme

In this experiment, catalysis of the decomposition of hydrogen peroxide by the enzyme catalase will be studied.

$$2 H_2O_2 \longrightarrow 2 H_2O + O_2$$

Hydrogen peroxide is produced in some cells and would be toxic if it accumulated. Catalase, which is found in red blood cells and in many plant cells, prevents the accumulation of hydrogen peroxide.

You will study the effect of temperature and pH on the catalytic power of catalase by measuring the rate of decomposition of hydrogen peroxide. In addition, the effect of $CuSO_4$, an inhibitor, will be examined. Finally, the enzyme will be denatured and its catalytic activity compared to that of the normal enzyme.

PROCEDURE

Before starting this experiment read carefully Sections H, "Glassworking" (pages 16-17) and "Using the Pipet" in Section J on page 18.

Carefully clean all equipment for this experiment, and rinse it with distilled water. Be careful not to contaminate the solution of the enzyme.

Measurement of Volume of Oxygen

The apparatus shown in Figure 38.1 will be used to determine the volume of oxygen. Plastic tubes are preferable to glass ones.

If glass tubes must be used, carefully follow the instructions on page 17.
Inserting glass tubing into rubber stoppers is hazardous if done improperly.

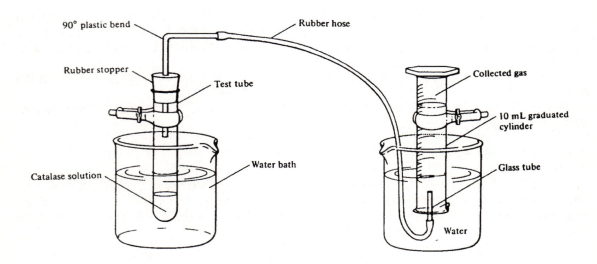

Figure 38-1. Apparatus to Collect Oxygen Gas.

The oxygen bubbled into the graduated cylinder displaces the water, and its volume is measured by the graduations on the cylinder. Have the apparatus properly secured so that the rubber stopper with its attached tubing may immediately be inserted in the test tube after mixing the enzyme and substrate.

The graduated cylinder should be filled with water and covered with your finger while inverting it into the beaker of water. The free end of the rubber tubing should be placed in the graduated cylinder prior to mixing the enzyme and substrate. Always use the same pipet for the enzyme solution. Use a different pipet for the hydrogen peroxide solution.

Part A—Effect of Temperature

Place the reaction test tube in a water bath containing ice water. Pipet 2.0 mL of 3% hydrogen peroxide into the test tube. Place a second test tube containing about 5 mL of catalase solution in the ice bath. After 5 minutes, pipet 2 mL of the catalase solution into the test tube containing hydrogen peroxide. Immediately stopper the test tube as shown in Figure 38-1, mix the contents by swirling, and note the time to the nearest second. Record on your Report Sheet the time required to produce 5.0 mL of gas. This time will vary with the activity of the enzyme sample. It is not necessary to collect exactly 5.0 mL of the oxygen gas. However, you should note both the volume and the time at which that volume is obtained. Record both the time and the volume of oxygen on the Report Sheet. Calculate the rate of oxygen production expressed in mL/sec.

Clean the reaction test tube thoroughly, and repeat the procedure with water at about 20°C. Both the enzyme and hydrogen peroxide solutions must be at the same temperature prior to mixing. Record the temperature, time of reaction, and volume of oxygen on your Report Sheet. Calculate the rate of oxygen production.

Clean the reaction test tube thoroughly, and repeat the procedure with water at 37°C. Record the temperature, time of reaction, volume of oxygen, and rate of oxygen production on your Report Sheet

Part B—Effect of pH

Pipet 2.0 mL of catalase solution into a clean reaction tube. Add 5 drops of 0.1 M HCl, place the tube in a water bath at about 20°C, and allow it to stand for about 5 minutes. Pipet 2.0 mL of 3% hydrogen peroxide into the tube and quickly stopper the tube. Record your observations on your Report Sheet.

Clean the reaction tube thoroughly, and perform a similar experiment with 5 drops of 0.1 M NaOH instead of HCl. Record your observations on your Report Sheet.

The pH of the acid and base solutions should be about 2 and 12, respectively. Compare your data with the result for the reaction without acid or base (pH = 7).

Part C—Effect of Inhibitor

Pipet 2.0 mL of catalase into a clean reaction tube, and add 4 drops of 0.1 M $CuSO_4$ solution. Place the test tube in a 20°C water bath, and allow it to stand for 5 minutes. Pipet 2 mL of 3% hydrogen peroxide into the test tube and quickly stopper the tube. Record your observations on the Report Sheet.

Catalysis by an Enzyme

Part D—Denaturation by Heat

Pipet 2.0 mL of catalase solution into the reaction tube, and heat it in a boiling water bath for 5 minutes. Cool the tube in water at room temperature, and place it in a 20°C water bath for 5 minutes. Pipet 2 mL of 3% hydrogen peroxide into the test tube and quickly stopper the tube. Record your observations on the Report Sheet.

Report Sheet

	Name		Instructor/Section		Date

Part A—Effect of Temperature

Temperature (°C)	Time (sec)	Volume of oxygen (mL)	Rate mL/sec
————	————	————	————
————	————	————	————
————	————	————	————

Part B—Effect of pH at 20°C

pH	Time (sec)	Volume of oxygen (mL)	Rate mL/sec
2	————	————	————
7	————	————	————
12	————	————	————

Part C—Effect of Inhibitor at 20°C

Inhibitor	Time (sec)	Volume of oxygen (mL)	Rate mL/sec
$CuSO_4$	————	————	————
None	————	————	————

Part D—Denaturation

State of Enzyme	Time (sec)	Volume of oxygen (mL)	Rate mL/sec
normal	————	————	————
denatured	————	————	————

Answer the questions on the following page.

Catalysis by an Enzyme

QUESTIONS

1. Explain the effect of temperature on the rate of the enzyme-catalyzed reaction of hydrogen peroxide in the range from 0 to 37°C.

2. How would the rate of the reaction be affected at 75°C?

3. What would be the shape of a graph of rate versus pH? Sketch this shape.

4. Suggest experiments that would show whether the copper(II) ion or the sulfate ion is the inhibitor of the enzyme-catalyzed reaction of hydrogen peroxide.

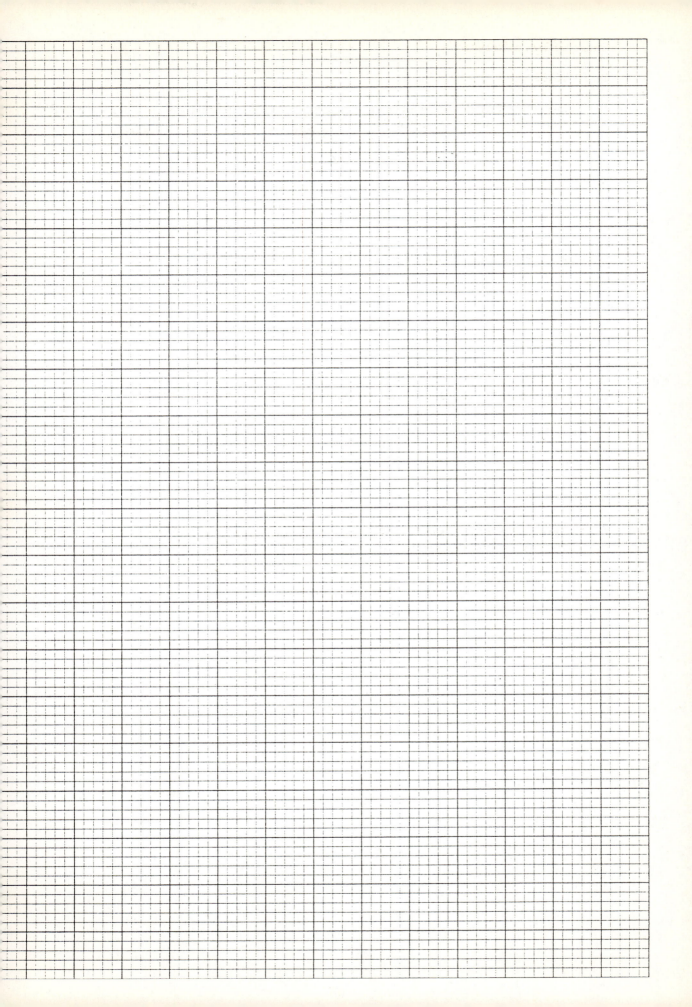

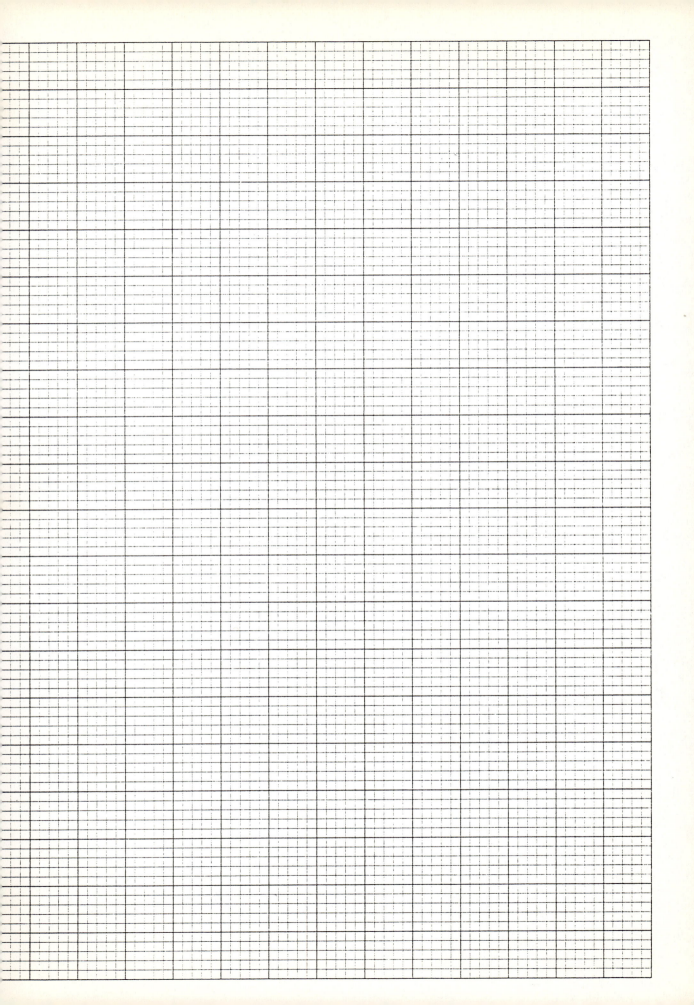

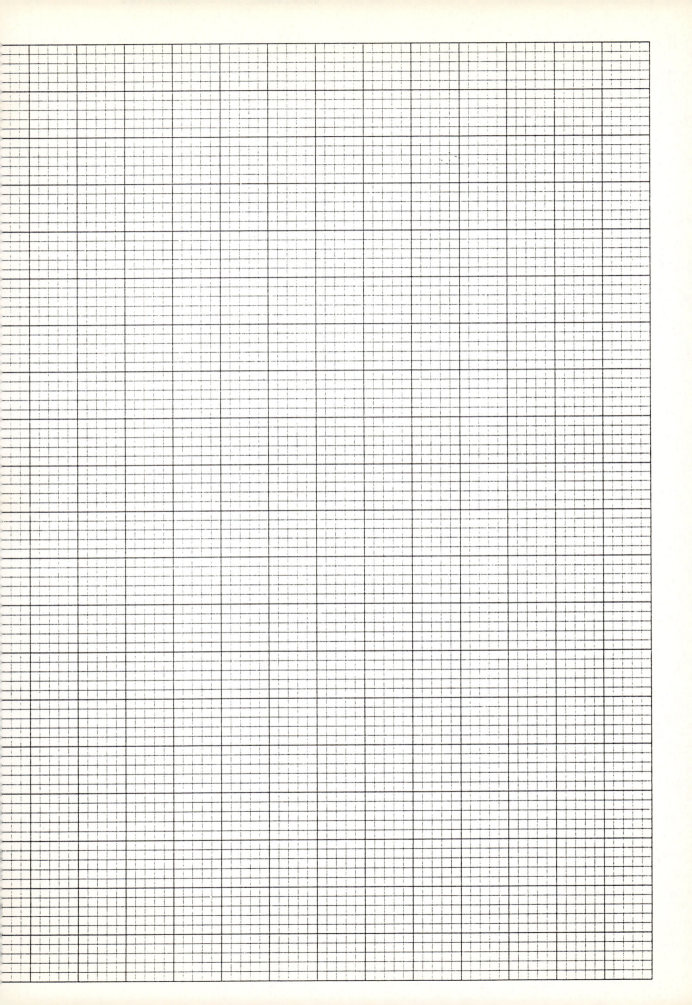

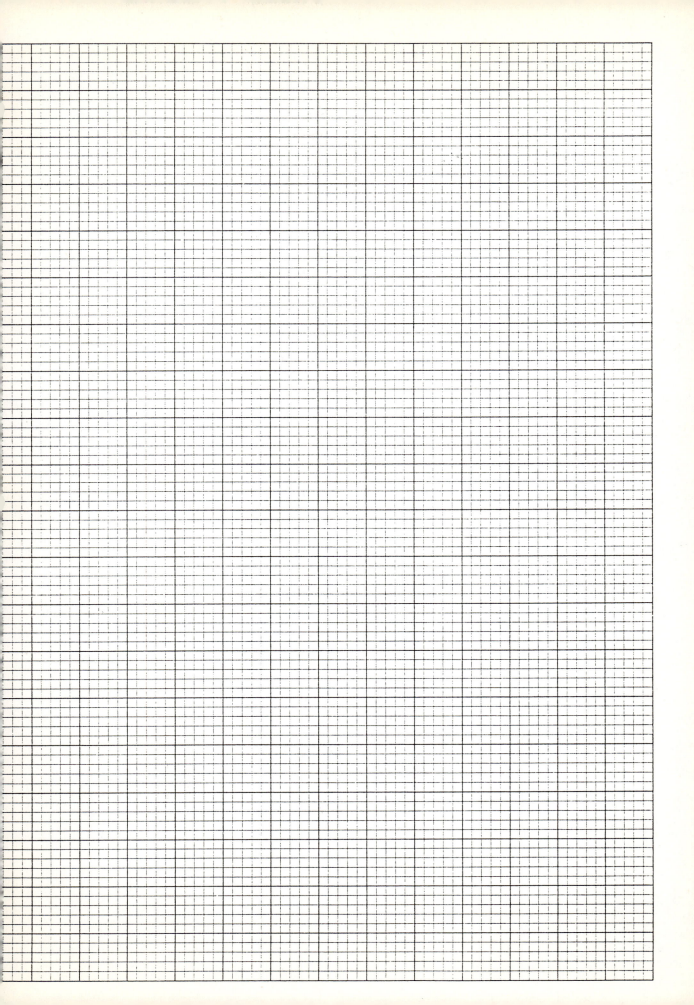

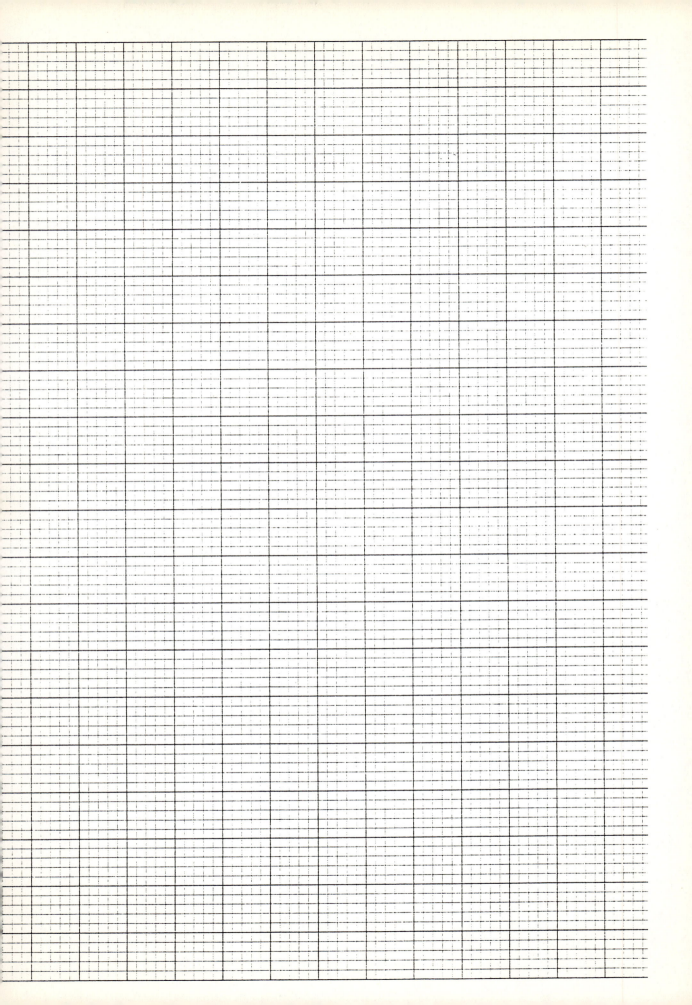

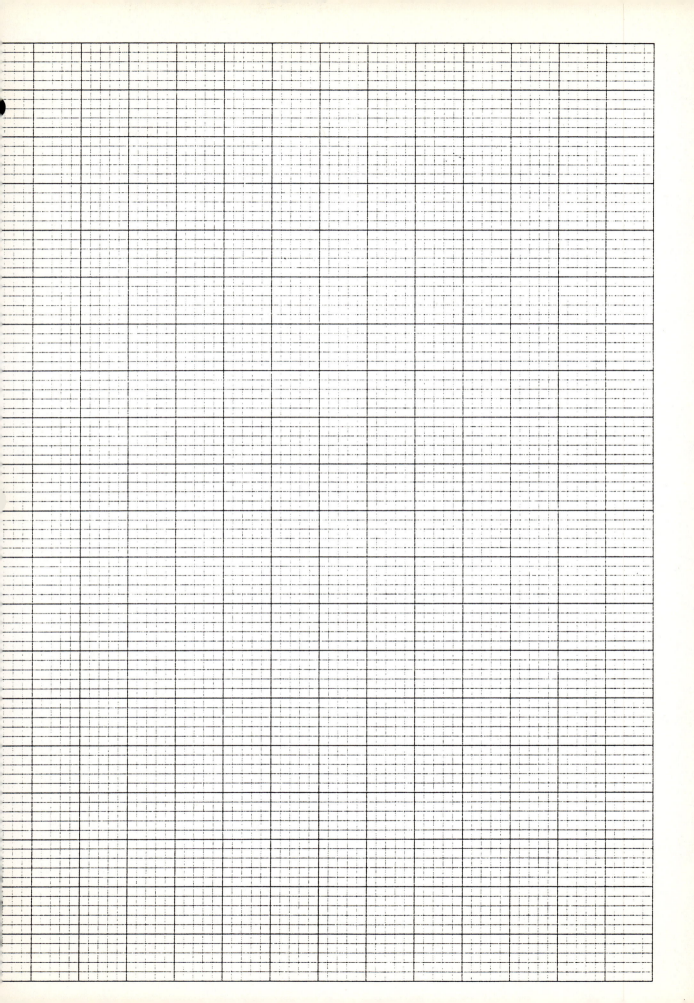

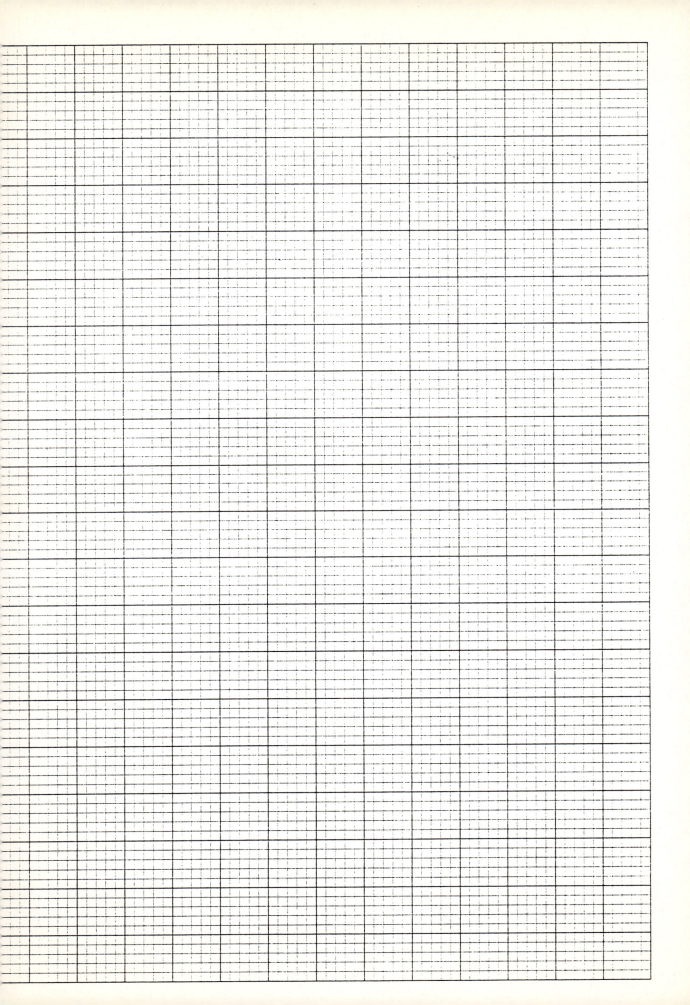

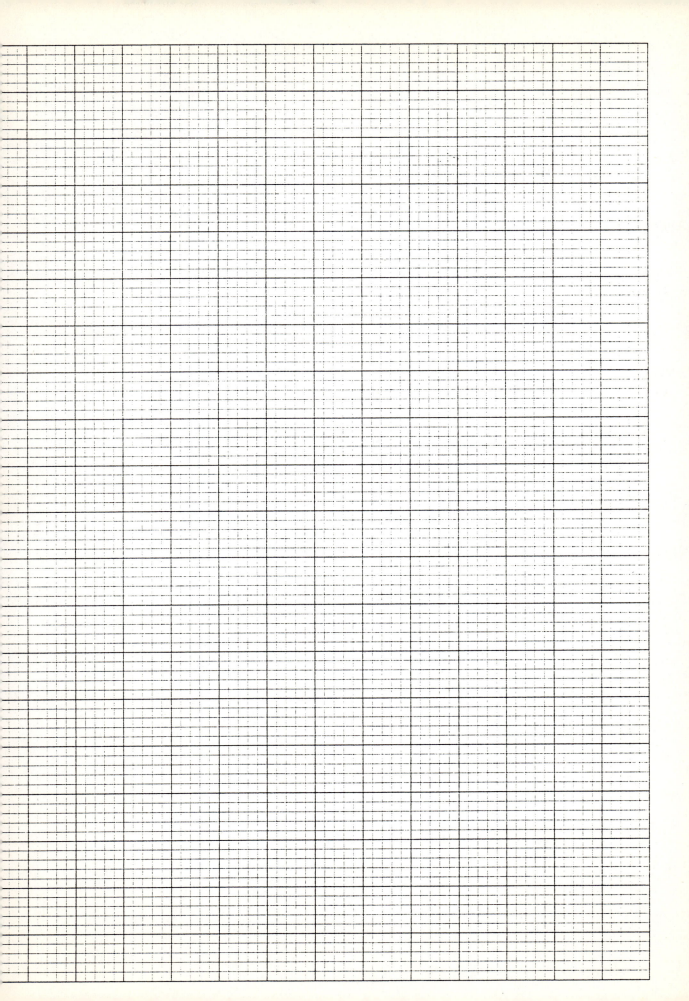